ISBN 978-0-483-01685-9
PIBN 10578637

This book is a reproduction of an important historical work. Forgotten Books uses state-of-the-art technology to digitally reconstruct the work, preserving the original format whilst repairing imperfections present in the aged copy. In rare cases, an imperfection in the original, such as a blemish or missing page, may be replicated in our edition. We do, however, repair the vast majority of imperfections successfully; any imperfections that remain are intentionally left to preserve the state of such historical works.

For support please visit www.forgottenbooks.com

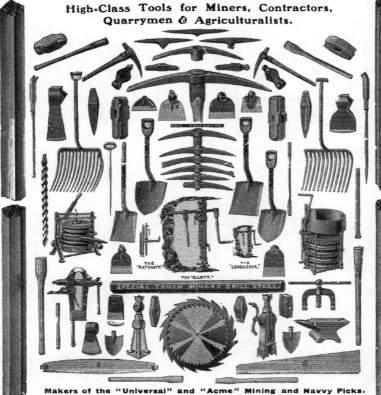
For contents of the Magazine for the month, see pages 2 and 4.  Index to Advertisers, pages 41, 43, 44, and 46.

# CONTENTS.

**No. 3.**    **MARCH, 1904.**    **Vol. IV.**

(Continued on Page 4.)

# CONTENTS.

*(Continued from Page 2.)*

# PAGE'S MAGAZINE

**An Illustrated Technical Monthly, dealing with the Engineering, Electrical, Shipbuilding, Iron & Steel, Mining, & Allied Industries.**

DAVIDGE PAGE, Editor,

Clun House, Surrey Street, Strand, London, W.C.

Telephone No.: 3349 GERRARD.
Telegraphic and Cable Address·
"SINEWY, LONDON."

## ANNOUNCEMENTS.

### Subscription Rates per Year.

**Great Britain**—In advance, 12s. for twelve months, post free. Sample copies, 1s. 4d., post free.

**Foreign and Colonial Subscriptions, 16s. for twelve months**, post free. Sample Copies, 1s. 6d., post free.

Remittances should be made payable to PAGE'S MAGAZINE, and may be forwarded by Cheque, Money Order, Draft, Post Office Order, or Registered Letter Cheques should be crossed "LONDON & COUNTY BANK, Covent Garden Branch." P O's and P.O O's to be made payable at East Strand Post Office, London, W.C. When a change of address is notified, both the new and old addresses should be given. All orders must be accompanied by remittance, and no subscription will be continued after expiration, unless by special arrangement Subscribers are requested to give information of any irregularity in receiving the Magazine

### Advertising Rates.

All inquiries regarding Advertisements should be directed to "THE ADVERTISEMENT MANAGER, Clun House, Surrey Street, Strand, London, W.C."

### Copy for Advertisements

should be forwarded on or before the 3rd of each month preceding date of publication.

**Editorial.**—*All communications intended for publication should be written on one side of the paper only, and addressed to "The Editor."*

*Any contributions offered, as likely to interest either home or foreign readers, dealing with the industries covered by the Magazine, should be accompanied by stamped and addressed envelope for the return of the MSS. i, rejected. When payment is desired this fact should be stated, and the full name and address of the writer should appear on the MSS.*

*The copyright of any article appearing is vested in the proprietors of PAGE'S MAGAZINE in the absence of any written agreement to the contrary.*

**Correspondence** *is invited from any person upon subjects of interest to the engineering community. In all cases this must be accompanied by full name and address of the writer, not necessarily for publication, but as a proof of good faith. No notice whatever can be taken of anonymous communications.*

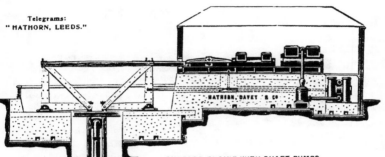

*SURFACE ENGINE WITH SHAFT PUMPS.*

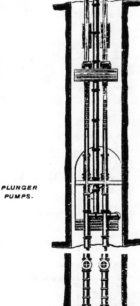

*PLUNGER PUMPS.*

*BUCKET LIFTS.*

# PUMPING MACHINERY

### Specialities—

## DIFFERENTIAL PUMPING ENGINES.
## ROTATIVE PUMPING ENGINES.

Horizontal and Vertical.
Compound and Triple.

---

## HYDRAULIC PUMPS.
## UNDERGROUND PUMPS.
## ELECTRIC PUMPS.
## WATER WORKS PLANT.

---

# HATHORN, DAVEY & CO.,
LIMITED,
# LEEDS,
# ENGLAND.

# BUYERS' DIRECTORY.

NOTE.—*The display advertisements of the firms mentioned under each heading can be found readily by reference to the Alphabetical Index to Advertisers on pages 41, 43, 44, & 46.*

*In order to assure fair treatment to advertisers, each firm is indexed under its leading speciality ONLY.*

*Advertisers who prefer, however, to be entered under two or more different sections can do so by an annual payment of 5s. for each additional section*

---

**Belting.**
Fleming, Birkby & Goodall, Ltd., West Grove, Halifax.
Rossendale Belting Co , Ltd.. 10, West Mosley Street, Manchester.

**Boilers.**
Clayton, Son & Co., Ltd., Leeds City Boiler Works, Leeds.
Galloways, Ltd., Manchester.
John Thompson, Wolverhampton.

**Boilers (Water-tube).**
Babcock & Wilcox, Ltd., Oriel House, Farringdon Street, London, E.C.
Cochran & Co. (Annan), Ltd., Annan, Scotland.
B. R Rowland & Co., Ltd., Climax Works, Reddish, Manchester.

**Bolts, Nuts, Rivets, etc.**
Herbert W. Perram, Ltd., Floodgate Street Works, Birmingham.
T. D. Robinson & Co , Ltd , Derby

**Books.**
E. & F. N. Spon, 125, Strand, London, W.C.

**Brass Engine and Boiler Fittings.**
Hunt & Mitton, Crown Brass Works, Oozells Street North, Birmingham.

**Cables.**
St. Helen's Cable Co., Ltd., Warrington, Lancashire.
Suddeutsche Kabelwerke A -G , Mannheim, Germany.

**Cement.**
J. H Sankey & Son, Essex Wharf, Canning Town, London, E.

**Clutches (Friction).**
David Bridge & Co., Castleton Ironworks, Rochdale, Lancashire.
H J. H King & Co , Nailsworth, Gloucestershire.

**Condensing Plant.**
Mirrlees-Watson & Co , Ltd , Glasgow.

**Consulting Engineers.**
G. H. Hughes, A M.I M.E., 97, Queen Victoria Street, London, E.C.

**Continental Railway Arrangements.**
South Eastern & Chatham Railway Co.

**Conveying and Elevating Machinery.**
Adolf Bleichert & Co., Leipzig-Gohlis, Germany.
Brown Hoisting Machinery Co., 39, Victoria Street, London, S.W.
Bullivant & Co., Ltd., 72, Mark Lane, London, E.C.
Fraser & Chalmers, Ltd , 3, London Wall Buildings, London, E.C.
Ropeways Syndicate, Ltd., 30, St. Mary Axe, London, E.C
Temperley Transporter Co , 72, Bishopsgate Street Within, London, E.C.

**Cranes, Travellers, Winches, etc.**
Joseph Booth & Bros. Ltd, Rodley, Leeds.
Thomas Broadbent & Sons, Ltd , Huddersfield.
Niles-Bement Pond Co , 23-25, Victoria Street, London, S W.

**Cranks.**
Clarke's Crank & Forge Co., Ltd , Lincoln, England.

**Cutters (Milling).**
E. G Wrigley & Co , Ltd., Foundry Lane Works, Soho, Birmingham.

**Destructors.**
Horsfall Destructor Co., Ltd., Lord Street Works, Whitehall Road, Leeds.

**Dredges and Excavators.**
Lobnitz & Co., Ltd., Renfrew.
Rose, Downs & Thompson, Ltd , Old Foundry, Hull.

**Economisers.**
E Green & Son, Ltd., Manchester.

**Electors (Pneumatic).**
Hughes & Lancaster, 47, Victoria Street, London, S.W.

**Electrical Apparatus.**
Allgemeine Elektricitäts Gesellschaft, Berlin, Germany.
Brush Electrical Engineering Co , Ltd., Victoria Works, Belvedere Road, London, S.E.
British Westinghouse Electric & Manufacturing Co., Ltd., Norfolk Street, Strand, London, W.C.
Crompton & Co., Ltd., Arc Works, Chelmsford.
Greenwood & Batley, Ltd.. Albion Works, Leeds.
The India Rubber, Gutta Percha, and Telegraph Works Co., Ltd , Silvertown, London, E.
Mather & Platt, Ltd., Salford Iron Works, Manchester.
Matthews & Yates, Ltd., Swinton, Manchester.
Newton Brothers, Full Street, Derby.
Phœnix Dynamo Manufacturing Co , Bradford, Yorks.
Simplex Steel Conduit Co , Ltd., 20, Bucklersbury, London, E.C.

**Electrical Apparatus** (*continued*).
SturteVant Engineering Co., Ltd , 147, Queen Victoria Street, London, E.C.
Turner, Atherton & Co., Ltd , Denton, Manchester.

**Engines (Electric Lighting).**
J & H McLaren, Midland Engine Works Leeds

**Engines (Locomotive).**
Baldwin Locomotive Works, Philadelphia, Pa., U.S.A.
Hunslet Engine Co , Ltd., Leeds, England.
Hudswell Clarke & Co , Ltd , Leeds, England.

**Engines (Stationary).**
Allis-Chalmers Co , 533, Salisbury House, Finsbury Circus, London, E.C.
Fraser & Chalmers, Ltd., 3. London Wall Buildings, London, E.C.
Robey & Co., Ltd., Globe Works, Lincoln, England.

**Engines (Traction).**
Jno Fowler & Co (Leeds), Ltd., Steam Plough Works, Leeds

**Engravers.**
Jno Swain & Son, Ltd., 58, Farringdon Street, London, E.C.

**Fans, Blowers.**
Davidson & Co , Ltd , "Sirocco" Engineering Works, Belfast, Ireland.
James Keith & Blackman Co , Ltd., 27, Farringdon Avenue London E.C.
Matthews & Yates, Ltd , Swinton, Manchester
The Standard Engineering Co., Ltd., Leicester.

**Feed Water Heaters.**
Royles, Ltd , Irlam, near Manchester.

**Fire Bricks.**
E J & J. Pearson, Ltd., Stourbridge.

**Firewood Machinery.**
M. Glover & Co., Patentees and Saw Mill Engineers, Leeds.

**Fountain Pens.**
Mabie, Todd & Bard, 93, Cheapside, London, E C

**Forging (Drop) Plants.**
Brett's Patent Lifter Co., Ltd., Coventry.

**Forgings (Drop).**
J H. Williams & Co , Brooklyn, New York, U.S.A.

**Furnaces.**
Deighton's Patent Flue & Tube Company, Vulcan Works. Pepper Road, Leeds.
Leeds Forge Co., Ltd , Leeds
W. F. Mason, Ltd , Engineers, Manchester.

**Gas Producers.**
W F. Mason, Ltd , Engineers, Manchester.

**Gears.**
William Asquith, Ltd., Highroad Well Works, Halifax.
Bifflotine Noiseless Gear Co , Levenshulme, Manchester.
E Arnold Pochin, Croff Street, Pendleton, Manchester.

**Gold Dredging Plant.**
Fraser & Chalmers, Ltd , 3, London Wall Buildings, London, E C

**Gauge Glasses.**
J. B Treasure & Co , Vauxhall Road, Liverpool.

**Hammers (Steam).**
Davis & Primrose. Leith Ironworks, Edinburgh
Niles-Bement Pond Co., 23-25, Victoria Street, London, S W.

**Hoisting Machinery.**
*See* Conveying Machinery.

**Horizontal Boring Machines.**
William Asquith, Ltd , Highroad Well Works, Halifax
Niles-Bement Pond Co., 23-25, Victoria Street, London, S W.

**Indicators.**
Dobbie McInnes, Ltd., 41 & 42, Clyde Place, Glasgow.

**Injectors.**
W. H. Willcox & Co , Ltd , 23, 34, & 36 Southwark Street, London.

**Iron and Steel.**
Askham Bros & Wilson, Ltd., Sheffield.
Brown, Bayley's Steel Works, Ltd., Sheffield.
Consett Iron Co., Ltd., Consett, Durham, and Newcastle-on-Tyne.
Farnley Iron Co., Ltd., Leeds England
Fried Krupp, Grusonwerk, Magdeburg-Buckau, Germany
Hadfield's Steel Foundry Co , Ltd , Sheffield.
J. Frederick Meiling, 14, Park Row, Leeds, England.
Parker Foundry Co., Derby.
Walter Scott, Ltd., Leeds Steel Works, Leeds, England.

**Laundry Machinery.**
W. Summerscales & Sons, Ltd., Engineers, Phœnix Foundry, Keighley, England.

**Lifts.**
Waygood & Co , Ltd., Falmouth Road, London, S.E.

**Lubricants.**
Blumann & Stern, Ltd , Plough Bridge, Deptford, London, S.E.
The Reliance Lubricating Oil Co , 19 & 20, Water Lane, Great Tower Street, London, E.C.

**Lubricators.**
Thomas A Ashton, Ltd., Norfolk Street, Sheffield.

**Machine Tools.**
George Addy & Co , Waverley Works, Sheffield.
William Asquith, Ltd., Highroad Well Works, Halifax, England.
Hy. Berry & Co., Ltd., Leeds.
Bertram's, Ltd., St Katherine's Works, Sciennes, Edinburgh.
Cunliffe & Croom, Ltd , Broughton Ironworks, Manchester.
Britannia Engineering Co , Ltd , Colchester, England
C. W. Burton Griffiths and Co , 1, 2, & 3, Ludgate Square, Ludgate Hill, London, E C.
Chas. Churchill & Co., Ltd , 9-15, Leonard Street, London, E C.
John Lang & Sons, Johnstone, near Glasgow
Luke & Spencer, Ltd , Broadheath, Manchester.
Niles-Bement Pond Co., 23-25, Victoria Street, London, S.W.
Northern Engineering Co , 1900, Ltd., King Cross, near Halifax
J. Parkinson & Son, Canal Ironworks Shipley, Yorkshire
Pratt & Whitney Co., 23-25, Victoria Street, London, S.W.
Rice & Co (Leeds), Ltd , Leeds, England.
Wm, Ryder, Ltd , Bolton, Lancs
G. F. Smith, Ltd , South Parade, Halifax.
Taylor and Challen, Ltd , Derwent Foundry, Constitution Hill, Birmingham
H. W. Ward & Co , Lionel Street, Birmingham.
T W. Ward, Albion Works, Sheffield
West Hydraulic Engineering Co., 23, College Hill, London, E.C.
Whitman & Barnes Manufacturing Co , 149, Queen Victoria Street, London, E.C.
Charles Winn & Co., St. Thomas Works, Birmingham

**Metals.**
Magnolia Anti-Friction Metal Co., Ltd., of Great Britain, 49, Queen Victoria Street, London, E.C.
Phosphor Bronze Co., Ltd., Southwark, London, S.E.

**Mircoscopes.**
W. Watson & Sons, 313, High Holborn, London, W.C.

**Mining Machinery.**
Chester, Edward, & Co , Ltd.
Fraser & Chalmers, Ltd , 3, London Wall Buildings, London, E C.
Hardy Patent Pick Co., Ltd., Sheffield.
Humbolt Engineering Co., Kalk, Nr. Cologne, Germany.
Ernest Scott & Mountain, Ltd , Electrical and General Engineers, Newcastle-on-Tyne, England.

**Office Appliances.**
"Business Engineer," c/o PAGE'S MAGAZINE, Clun House, Surrey Street Strand, London, W.C.
Library Bureau, Ltd., 10, Bloomsbury Street, London, W.C.
Library Supply Co , Bridge House, 181, Queen Victoria Street, London, E.C.
Lyle Co., Ltd., Harrison Street, Gray's Inn Road, London, W.C.
Partridge & Cooper, Ltd , 191-192 Fleet Street, London, E C
Rockwell-Wabash Co. Ltd., 69, Milton Street, London, E C.
Shannon, Ltd., Ropemaker Street, London, E.C.
The Trading and Manufacturing Co., Ltd , Temple Bar House, Fleet Street, London, E C.

**Oil Filters.**
Vacuum Oil Co , Ltd , Norfolk Street, London, W.C

**Packing.**
Lancaster & Tonge, Ltd., Pendleton, Manchester.
The Quaker City Rubber Co , 101, Leadenhall Street, London, E C
United Kingdom Self-Adjusting Anti-Friction Metallic Packing Syndicate, 14, Cook Street, Liverpool
United States Metallic Packing Co., Ltd., Bradford,
J. Bennett von der Heyde, 6, Brown Street, Manchester.

**Paper.**
Lepard & Smiths, Ltd., 29, King Street, Covent Garden, London, W C.

**Photo Copying Frames.**
J. Halden & Co., 8, Albert Square, Manchester

**Photographic Apparatus.**
Marion & Co., Ltd , 22, 23, Soho Square, London, W.
W. Watson & Sons, 313, High Holborn, London, W.C.

**Pistons.**
Lancaster & Tonge, Ltd , Pendleton, Manchester.

**Porcelain.**
Gustav Richter, Charlottenburg, near Berlin, Germany

**Presses (Hydraulic).**
Niles-Bement Pond Co., 23-25, Victoria Street, London, S.W.

**Printing.**
Southwood, Smith & Co., Ltd., Plough Court, Fetter Lane, London, E.C.

**Publishers.**
Anglo-Japanese Gazette 39, Seething Lane, London, E.C.
Association de la Presse Technique, 20, Rue de la Chancellerie, Brussels, Belgium.
Electroteknisk Tidsskrift Publishing Co., Copenhagen, N, Sweden

**Publishers** (*continued*).
Charles Griffin & Co., Ltd., Exeter Street, Strand, London, W.C.
New Zealand Mines Record, Wellington, New Zealand.
Shipping World, Ltd , Effingham House. Arundel Street, Strand, London W.C.
South African Mines, Commerce, and Industries. Johannesburg.

**Pulleys.**
John Jardine, Deering Street, Nottingham.
H J. H. King & Co , Nailsworth, Glos

**Pumps and Pumping Machinery.**
Blake & Knowles Steam Pump Works, Ltd., 179, Queen Victoria Street, London, E C.
Drum Engineering Co , 27, Charles Street, Bradford.
Fraser & Chalmers, Ltd., 3, London Wall Buildings, London, E.C.
J. P Hall & Sons, Ltd., Engineers, Peterborough.
Hathorn, Davey & Co., Ltd., Leeds, England
Tangyes, Ltd , Cornwall Works, Birmingham

**Radial Drilling Machines.**
William Asquith, Ltd , Highroad Well Works, Halifax.
Niles-Bement Pond Co , 23-25, Victoria Street, London, S W.

**Rails.**
Wm. Firth, Ltd , Leeds.

**Railway Wagons.**
W. R. Renshaw & Co , Ltd , Phœnix Works, Stoke-on-Trent.

**Riveted Work.**
F. A Keep, Juxon & Co , Forward Works, Barn Street, Birmingham

**Roof Glazing.**
Mellowes & Co , Sheffield.

**Roofs.**
D. Anderson & Son, Ltd., Lagan Felt Works Belfast.
Alex Findlay & Co . Ltd , Motherwell, N B.
Head, Wrightson & Co., Ltd , Thornaby-on-Tees

**Scientific Instruments.**
Cambridge Scientific Instrument Co., Ltd., Cambridge.

**Stampings.**
Thos Smith's Stamping Works, Ltd , Coventry
Thomas Smith & Son of Saltley, Ltd , Birmingham.

**Steam Traps.**
British Steam Specialties, Ltd., Fleet Street, Leicester.
Lancaster & Tonge, Ltd , Pendleton, Manchester.

**Steel Tools.**
Saml Buckley, St Paul's Square, Birmingham
Pratt & Whitney Co., 23-25, Victoria Street, London, S W.

**Stenotypers.**
Stenotyper (1902), Ltd., 25, Southampton Row, London, W.C.

**Stokers.**
Ed Bennis & Co., Ltd , Bolton, Lancs
Meldrum Brothers, Ltd , Atlantic Works, Manchester.

**Stone Breakers.**
S. Pegg & Son, Alexander Street, Leicester.

**Time Recorders.**
Howard Bros , 10, St. George's Crescent, Liverpool, and 100c, Queen Victoria Street, London, E C
International Time Recording Co , 171, Queen Victoria Street, London, E.C.

**Tubes.**
Thomas Piggott & Co , Ltd , Spring Hill, Birmingham.
Tubes, Ltd., Birmingham.
Weldless Steel Tube Co., Ltd., Ickmeld Port Road, Birmingham.

**Turbines.**
G Gilkes & Co., Ltd , Kendal.
W Gunther & Sons, Central Works Oldham.
S. Howes, 64, Mark Lane, London, E C

**Typewriters.**
Empire Typewriter Co., 77, Queen Victoria Street, London, E.C.
Oliver Typewriter Co., Ltd , 75, Queen Victoria Street, London, E.C.
Yost Typewriter Co., 50, Holborn Viaduct, London, E C.

**Valves.**
Alley & MacLellan, Ltd., Glasgow.
Scotch and Irish Oxygen Co., Ltd , Rosehill Works, Glasgow.

**Ventilating Appliances.**
Matthews & Yates, Ltd., Swinton, Manchester.

**Wagons—Steam.**
Thornycroft Steam Wagon Co., Ltd., Homefield Chiswick, London, W.

**Weighing Apparatus.**
W. T. Avery & Co., Soho Foundry, Birmingham, England.
Samuel Denison & Son, Hunslet Moor, near Leeds.

**Wells Light.**
A. C. Wells & Co , 100A, Midland Road, St Pancras, London, N.W.

**"Woodite."**
"Woodite" Company, Mitcham, Surrey

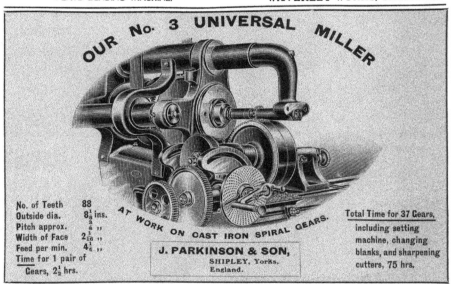

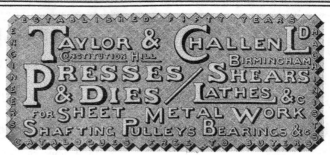

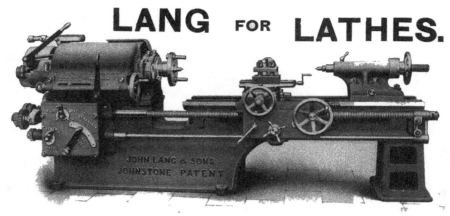

22

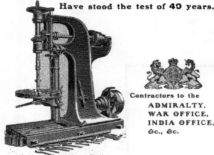

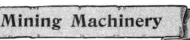

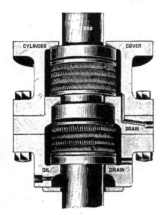

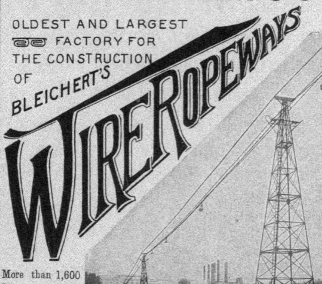

31

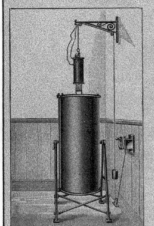

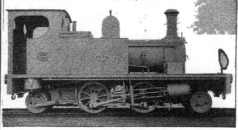

Our Expansion Vertical
Files are particularly .
well adapted for Filing
Letters, Catalogues, and
Invoices.

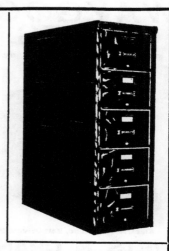

SEND FOR OUR . .

# VERTICAL FILE

CATALOGUE No. 100,
Which is Fully Descriptive.

ROCKWELL-
WABASH .
Co., Ltd., . .
69, Milton St.,
London, E.C.;
50, Deansgate
Arcade, . .
Manchester .

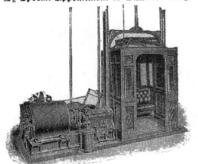
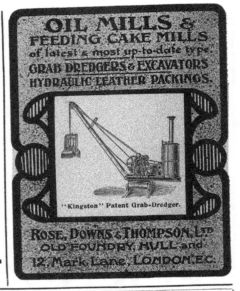

# INDEX·TO·ADVERTISERS

# MATTHEWS & YATES, Ltd., Swinton, Manchester.

## FANS FOR WATER COOLING TOWERS, &c.

# Index to Advertisers—(Contd.)

# Index to Advertisers—(Contd.)

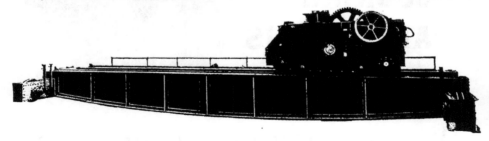

# Index to Advertisers—(Contd.)

# MAGNOLIA METAL..

**Best Anti-Friction Metal for all Machinery Bearings.**

"Flower" Brand.

"Flower" Brand.

The Name and Trade Mark appear on each Box and Ingot.

## Magnolia Anti-Friction Metal Company, of Great Britain, Limited,

49, QUEEN VICTORIA STREET,

### LONDON, E.C.

Telephone : 5925 Bank.    Telegrams : "MAGNOLIER, LONDON."

BERLIN : FRIEDRICH STRASSE, 71.    PARIS : 50, RUE TAITBOUT.
LIEGE, BELGIUM : 36, RUE DE L'UNIVERSITE.
GENOA, VIA SOTTORIPA : I, PIANO NOBILE.

47

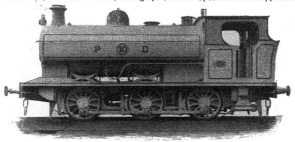

Specially taken for PAGE'S MAGAZINE.]

MR. JOHN GAVEY, C.B., M.INST.C.E., M.I.E.E., ENGINEER-IN-CHIEF AND ELECTRICIAN TO THE BRITISH POST OFFICE.

# PAGE'S MAGAZINE

An Illustrated Technical Monthly, dealing with the Engineering, Electrical, Shipbuilding, Iron and Steel, Mining and Allied Industries.

VOL. IV.                    LONDON, MARCH, 1904.                    No. 3.

FIG. 21.   GENERAL VIEW OF WORKMEN'S VILLAGE, SCHLAEGEL AND EISEN COLLIERY.

# NOTES ON THE WESTPHALIAN COAL FIELD.

BY

DAVID A. LOUIS, F.I.C., M.I.M.E., M.A.I.M.E., F.C.S.

IV.

IN previous numbers of this Magazine the writer considered the topography, geology, and the character of the coal seams of the Westphalian coal field, also the conditions incident to a production of 60,000,000 tons from that field, so far as related to the methods of working, means of transport, arrangements for winding, and appliances for draining; leaving such matters as ventilation and incidental factors; the preparation of the produce for market and its disposal and business and labour considerations for treatment in the present article. The regulated output has since been raised to 65,000,000 tons a year.

### VENTILATION.

Under this head will be considered: The natural atmosphere of the mines, the inherent temperature, and the means employed to control these inconveniences.

The atmosphere encountered in the West-phalian mines is decidedly bad and gassy. The gassy character is affected by two circumstances, the position of the coal in the field and the character of the coal being worked. Those mines working out-crop coal, indicated at A on the map on page 484, vol. III., were much freer from gas than those mines working coal under the marly over-burden indicated at C on the same map. In fact, out of seventy-eight out-crop mines only six yielded more than 140 cubic feet of gas per ton of coal, and these were not only situated in proximity to the marly over-burden, but they were working the most gassy coal seams. Whereas of 132 mines working coal under the marls, 72 yielded 140 cubic feet and upwards per ton of coal. But more noteworthy still is the effect of the character of the coal being worked on the atmosphere of the mines; the fat bituminous

FIG. 17.  MINERS PROTECTED BY THE STOLZ SMOKE MASK.

coal seams yielding most gas, the gas and flaming coals coming next and the lean coals emitting the least.

Of 55 mines working in the latter seams, 30 gave below 17 cubic feet of gas per ton of coal, and none gave over 105 cubic feet; in the fat bituminous coal seams there were 119 mines working, and of these only 8 gave less than 17 cubic feet, quite a number gave quantities varying from 70 to 420 cubic feet, several gave even more. Amongst the highest were Gneisenau with 1,400, Hibernia with 1,800, Grimberg with 1,977, and Dahlbusch II. with 2,380 cubic feet of gas per ton of coal mined. Turning to the flaming and gas coal seams of 36 mines, 6 gave less than 17 cubic feet, 25 gave quantities varying from 17 to 420 cubic feet, and 5 yielded volumes between 420 and 700. Blowers are not infrequent and the gas is utilised in many mines; at Hansa Colliery, for instance, to light the pit eye, and at Consolidation Colliery for warming the compressed air where used for power purposes.

The Westphalian mines are amply provided with airways. Some very gassy mines are even provided with double airways, and in those subject to blowers special provision is made for driving this excess of gas direct into the return airway. Only 10 mines had less than 1,100 yards of airways, whilst 5 had more than 8,750 yards, the remaining mines had lengths of airways intermediate between these extremes. Air splitting, so as to give different parts of the mines their own supply of fresh air is, of course, in vogue, and the air currents are conducted to the working faces by various means to suit circumstances: By dividing walls, by masonry conduits, by air-boxes, by brattice, by special fans, by compressed air either employed in working drills, etc., or specially provided for ventilative purposes. There were some 434 ventilating shafts on the field, 43 were divided to serve both for up—and down—cast, 190 served for downcast only and the rest for upcast only; of the latter many were in use for drawing coals, and in those instances various contrivances were in use to prevent leakage of air, such as: Briart's shaft cover; air-locks with double doors and the Neumühl arrangement referred to in the January number, pages 38-41.

The currents were generated by fans; of old types the Capell, of new ones the Rateau fan being mostly in use, and these were being driven in some cases by electricity but mostly by steam engines. As a matter of course the quantity of air required varies very considerably in different mines, but the object is to keep the gas below one per cent. in the returns; in fact no mine exceeded this, and by far the preponderating number kept the gas content of the returns well below half that amount. The speed of the currents did not generally exceed 20 ft. a second, nevertheless a good many exceeded this rate and 28 even maintained a rate of over 26 ft. The equivalent orifice varied from under 1 square yard to above 4½.

In Table I. are set forth numbers showing the extremes relating to the air supply in different mines in 1900.

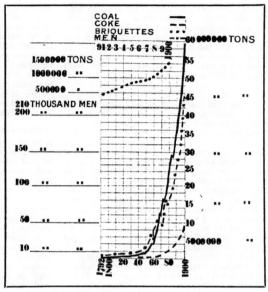

FIG. 18.   OUTPUT DIAGRAM.

## TABLE I.

AIR CIRCULATION IN WESTPHALIAN MINES.

| Cub. ft. air per min. | No. of mines. | Cub. ft. air per man per min. | No. of mines. | Cub. ft. air per ton drawn. | No. of mines. |
|---|---|---|---|---|---|
| Below 17,560 | 22 | Below 70 ... | 0 | Below 35 .. | 10 |
| Above 211,800 | 3 | Above 350 .. | 4 | Above 280 .. | 2 |
| Intermediate | 185 | Intermediate | 206 | Intermediate | 198 |

The natural temperature of the mines in Westphalia increases as is the case elsewhere with the depth, and the result of observations is set forth in Table II. The increase, on an average, being 1 deg. F. for each additional 8 fathoms. This temperature, of course, has to be kept down, and it is kept down by ventilation. In many cases where the heat greatly impeded work, improved ventilation has rectified the inconvenience; a good instance of this occurred in Hansa Colliery, where a temperature of 84 deg. F. existed, and it was only possible to work six-hour shifts, but by judicious air-splitting the temperature

was reduced to 76 deg., and work could be endured for the normal number of hours.

## TABLE II.

TEMPERATURE IN WESTPHALIAN MINES.

| Depth in fathoms | Degrees Fahrenheit. |
|---|---|
| 108 | 59 |
| 164 | 65 |
| 218 | 72 |
| 273 | 78 |
| 328 | 85 |
| 383 | 91 |
| 437 | 98 |
| 492 | 104 |
| 547 | 111 |

### DUST.

This is a point that those who mine in Westphalia dare not overlook. The tendency to produce dust is influenced to a certain extent by the caking character of the coals. The fat coals are more prone to form dust than the gas and flaming coals. The lean coals can also give rise to plenty of dust, but, generally, these seams have sufficient natural moisture to keep the dust down. Inclined seams give more dust in working than flat seams. The dust from any of the seams can cause explosion if fine enough, but the dust from the fat coals is the most dangerous. It is therefore realised in Westphalia that the dust must be dealt with wherever coal is mined.

### SPRINKLING.

To keep the dust down a considerable amount of sprinkling is therefore done in the mines, and this becomes quite a formidable item of expenditure as well for the maintenance of the system of pipes itself as for repairing damage done to roads by the water. In 1900 there were 2,500 miles of sprinkling piping under ground, and in one mine alone, Consolidation, there were 79 miles of pipe which cost £18,000. The cost of sprinkling varied from $8\frac{3}{4}$d. to $21\frac{1}{4}$d. per ton of coal mined for the year; the great difference between the lowest and highest, of course, was due to more renewals, etc., in one case than in others, an average cost for sprinkling was between 11d. and 14d. per annum per ton of coal mined.

The quantity of water used for this purpose, the special arrangements and the men employed in the sprinkling are quite noteworthy. In one of Bergrath Behrens' mines, Hibernia, where 850 tons of coal were drawn a day, 1,000 cubic feet of water were used in sprinkling all open roads and 126 working places. The roads were sprayed by special men, giving employment to ten men in the morning shift and three in the afternoon, no road spraying being done in the night shift; and the spraying in the working places being done by the miners themselves. Smiths were also employed constantly at repairs. Mine water was, in the Hibernia mine, used for spraying, and was taken off from the pumps at a convenient level, where it was stored in a tank and gave a water column equivalent to 27 atmospheres pressure in the shaft, which became from 8 to 18 atmospheres at points of application. The water was distributed by galvanised wrought-iron pipes tested to 60 atmospheres, and there were 524 yards of piping $3\frac{1}{4}$ in. internal diameter, 2,050 yards 2 in., and 53,270 yards of 1 in. There were 1,169 hydrants connected to the pipes at intervals of 160 ft. apart along the roads, a distance to be reduced, inasmuch as it necessitated 60 to 80 ft. of flexible rubber hose, which has to be stout enough to resist the pressure and such a length is therefore very heavy for the sprinklers to drag about the roads; shorter lengths were in use at the working faces; a jet

was used at the end of each piece of hose. The system of piping was provided with 574 stopcocks, so as to shut off any part if not required, or for repairs or for lengthening. The cost of the installation and material for this sprinkling plant was £68,125, whilst over £1,171 a year was being spent on its supervision and maintenance. In another mine 1,900 cubic feet of water was used with six master sprinklers and six smiths at work in the morning and afternoon shift, and two sprinklers in the night shift for an output of 1,000 tons a day.

### EXPLOSIVES, EXPLOSIONS AND SAFETY LAMPS

have such a great influence on the safety of mining operations that a word or two must be said about them. In 1900 safety explosives had replaced others in Westphalia, except in a few mines working the lean coal seams and the gas coal seams where black powder was still in use. Fuses, even safety fuses, were being replaced by electric firing. For lighting the pit bottom and stations incandescent electric lamps were generally employed, portable electric lamps, except in connection with rescue appliances, were not in use. Single-gauze safety lamps, of the Marsant type. with re-lighting arrangement, made by Friemann and Wolf, to burn benzine were the most prevalent working lamps. It is satisfactory to note that at the time of the writer's visit to the district, although the output had been increasing, there was a diminution in the number of explosions as well as in the number of fatalities arising therefrom; this is attributable to many of the precautionary measures already noted: improved ventilation,

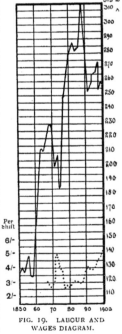

FIG. 19. LABOUR AND
WAGES DIAGRAM.

sprinkling, safety explosives and electric firing. It is noteworthy that most elaborate appliances are always kept in readiness for rescue work, for protection from fire and for combatting fire ; men, moreover, are constantly kept drilled in the use of these appliances. These departments, in fact, form quite remarkable features at most of the mines and make quite striking displays, the medical department with operating tables, medicine chests and surgical appliauces ; the rescue department with respiratory apparatus and special clothes to enable men to go into smoky or other bad atmospheres, stretchers, etc. ; these were particularly good at Shamrock mine, another of those under Bergrath Behrens' management. In fact, there was an excellent exhibit of this description from this mine at the Düsseldorf Exhibition ; moreover, the writer had an opportunity of witnessing the thoroughness with which the practice is carried out. Men appropriately clothed and equipped with pneumatophors were sent into a room hermetically closed, but provided with windows for inspection, and with a pernicious smoky atmosphere to make the practice realistic. The practice consisted in the search for injured and other companions, their treatment and removal to a place of safety. The extinction of fire underground, the removal of debris and the setting of timbers and restoration of order in the disturbed working. The type of pneumatophor used in the case referred to was provided with two cylinders of oxygen carried low down on each man's back and a respiratory sack in front into which a solution of caustic soda had been tipped to absorb the products of respiration. Two cylinders of oxygen were used, so as to warn the man when one gives out that half his store was exhausted. Fig. 17 shows· men protected by the Stolz smoke mask building up a place in a bad atmosphere. The fire-brigade department, with their showy appliances kept in perfect order, form distinctly attractive looking adjuncts to many a colliery in the neighbourhood.

FIG. 20.  WASHING AND CHANGING HALL, SHAMROCK COLLIERY.

## PREPARING THE COAL FOR MARKET.

Only a very small proportion of the coal raised in Westphalia can be profitably placed on the market directly as it comes from the mines. It is therefore subjected to some sort of preparatory treatment either to improve the value of good stuff or to render the useless useful ; the means employed are screening and picking or the more elaborate washing, and much of the output in the form of finely divided coal is subsequently coked or made into briquettes. The washing plants used are mostly either those of Humboldt of Kalk, near Cologne, or of Schüchtermann and Kremer of Dortmund or of Baum of Herne. The latest installations of the last-named maker were being arranged to wash the coal without previously separating into different sizes. as had hitherto been the practice, the washed coal being sized subsequently. Such an arrangement would require fewer jigs, less water, and less power, and would entail a simpler and less expensive plant.

About 9¾ million tons out of the sixty million tons of coal mined was recovered in the form of mud from the washeries and was subsequently converted into coke and briquettes.

*Coke making* is an industry that has been showing steady progress in Westphalia. All stages of the practice have been passed through : heaps,  mounds, stacks and various forms of the bee-hive ovens, but all these have long been discarded and chamber ovens are now exclusively used. The prevailing type of oven was the Otto-Coppée and in various forms some six thousand of them took part in the production of the millions of tons of coke in 1900. But other systems have also been introduced, including those of Collin, von Bauer, Brunck, Koppers, etc. An important factor in connection with the coke ovens is the recovery of by-products, a factor which is not neglected in the district under notice so that in the year, 1900, 12,000 tons of benzol were recovered from the by-products, and more than 77,000 tons of tar and a little over 36½ thousand tons of sulphate of ammonia. Besides these 1,530,000 tons of briquettes were produced in ninety-two presses, for this purpose only the fines from washing the leaner or semi-bituminous coal mixed with pitch were used.

#### CHARACTER OF THE PRODUCE.

Thirty per cent. of the output of the field was gas and flaming coal, fifty-seven per cent. fat and bituminous coal and thirteen per cent. lean coal.

The gas coal and the flaming coal burn with long flames and are comparatively strong and hard, which renders them suitable for storage and transport. The gas coal, as the name signifies, contains much gas, and is almost exclusively used for gas-making. The flaming coal is suitable for any purpose where a good flame is wanted; in the domestic hearth and in the puddling furnace for example. The fat coal also burns with a long flame and is a good heat producer, but it is particularly distinguished by its great caking properties which renders it eminently suitable for coke making. It is also in demand for many other purposes. The lean coal burns with less smoke and soot than the fat coal, and therefore is suitable for use in textile and other industries where smoke is undesirable; the fine coal produced in the washing operation in this case is that used in the manufacture of briquettes. These briquettes stand storage well, the pitch protecting them from absorbing moisture; moreover, their rectangular shape and uniform size render them good for transport. They are largely used on locomotives, steam ships, in dwelling houses, and are replacing flaming coal in iron works, sugar factories, glass works and porcelain and cement mills. Over 13¾ million tons of fine coal was used in the production of coke and briquettes.

Fig. 18 shows the development of the coal, coke, and briquette industry in the Westphalian field from the earliest time up to the commencement of this century. In the figure vertical lines refer to the years, the horizontal lines to quantities. The tonnages to the right of the figure relate to coal and coke, those to the left of it to briquettes, on this side the numbers relating to men employed are shown. We see in the early days the production was quite small, it was less than two million tons, and employing less than 10,000 men up to 1840; but it progressed steadily until 1850, when the iron industry awakened into activity, and this, with the extension of railways, exerted a marked effect on the Westphalian coal output, which since 1860 has continued to increase vigorously, until in the year more particularly referred to in these notes, it amounted to nearly sixty million tons of coal and 9⅔ million tons of coke and considerably over 1½ million tons of briquettes; and that the number of workmen employed approached 228,000. It will be noticed how the labour curve varies with the output; moreover, that coke first appears about 1841 and briquettes only fifty years afterwards, that is, in 1891.

#### DESTINATION OF THE PRODUCE.

Without going into details of the different localities, Table III. sets forth the destination

TABLE III.

WESTPHALIAN EXPORTS.

| District. | Million Tons. | | Briquettes in 100,000 tons. |
|---|---|---|---|
| | Coal. | Coke. | |
| Neighbourhood ... . | 29⅔ | 3⅑ | 7 7/9 |
| Elsewhere in Germany... | 9⅓ | 3⅓ | 7 2/9 |
| Abroad ... ... ... | 4 | 2⅑ | ¼ |
| Total sent from Collieries, etc. ... ... | 43 | 8 5/9 | 15¼ |

of the produce for the year concerned. Most of this went away by rail but about seven million tons were carried on the Rhein from the ports of Ruhrort, Duisberg and Hochfeld, whilst 54,000 tons found their way down the Dortmund Ems Canal.

The business departments of these great mines are naturally not without interest. It is well to remember that up to 1851 they were entirely under the control of the State; the Mining Office decided which collieries should be worked and what production each should yield and fixed the selling price. Want of railway communication at that time restricted the market within quite narrow limits, and hence the smallness of the consumption indicated in fig. 18 is readily explained. But in 1851 a new law came into operation, allowing private parties to work the mines, and so although there were only 205 mines in 1854, 298 were at work in 1857. The output, which was worth £500,000 in 1851, was worth £1,640,000 in 1858, whilst after the Franco-German War the value of the output rose considerably, reaching £7,900,000 in 1874, and prices had risen from 5s. 3d. a ton in 1869 to 11s. a ton. This state of things did not continue, and in 1879 the price had got back to 4¼s. per ton. So the district has had its periods of prosperity but has also had periods of adversity, when many mines were carried on at a loss. Of late years three organisations have handled the produce of the district, and fixed the output of each of the collieries concerned, the " Rheinisch-Westfälisches Kohlensyndikat " dealt with 87·4 per cent. of the entire output of the field, the rest of the output was mostly derived from collieries belonging to one or other of the great iron companies who consumed it themselves. The " Westfälische Koks-syndikat " in the same way dealt with 98·5 per cent. of the coke produced, whilst the third syndicate disposed of the whole output of briquettes. This arrangement insures a certain regularity in the prices obtained, and better prices which are constantly improving have ruled since the syndicates have had the matter in hand, and many of the mines have been able to pay magnificent dividends. The cost of production is dependent on various circumstances, and, therefore, the actual cost of the coal at pit mouth is different at different collieries, but the average was about 6s. or 7s.

The collieries are worked some by companies and some by corporations. Twenty-nine of the companies, representing a capital of £22,807,600, and paying £5,696.950 per annum rent, etc., each produced over 200,000 tons of coal in 1900, or, altogether, 34.724,000 tons. The smallest of these worked an area of 772 acres with a capital of £150.000 ; the greatest of these companies worked an area of 41,688 acres with a capital of £2,700,000, and produced 5,460,000 tons, this is the " Gelsenkirchener Bergwerks-

Aktiengesellschaft." The next largest outputs, 5,187,000 and 3,423,000 tons, respectively, were those of the " Harpener Bergbau-Aktiengesellschaft " and of " Hibernia," working, the one 30,704 acres with a capital of £2,600,000. the other 16,022 acres with a capital of £1,890,000. The " Harpener Bergbau-Aktiengesellschaft " worked seventeen pits with 20,500 workmen and 715 officials great and small. The largest corporation property was the Zollverein, working an area of 5.407 acres and putting out 1,753.000 tons of coal ; whilst the smallest of these areas producing over 200,000 tons was 254 acres.

## LABOUR.

The relation of the labour employed to the output is set forth in fig. 18, and shows with some clearness that the number of men employed increases with the greater output. Naturally enough, as the prosperity of the district grew more rapidly than the indigent population a great number of men are drawn from outside, mostly coming from Prussian Polish provinces, but others coming from various foreign countries, many not being able to speak German, but no important work involving risk is entrusted to any who do not speak and understand German. People over sixty or lads under sixteen are not allowed to work underground. The underground shifts are eight hours each, but on the surface eight to twelve hour shifts are worked, exclusive of time occupied in reaching and leaving work ; fig. 19 shows the hewers' earnings per shift and also tonnage per man per annum. In this figure, again, the vertical lines indicate years, the horizontal quantities. The tonnage per man per year in other German coal districts in 1900 was 346 in Upper Silesia, 213 in Lower Silesia, 227 in the Saar

FIG. 22. STORE AND A PORTION OF THE VILLAGE, SCHLAEGEL AND EISEN COLLIERY.

district, so Westphalia holds its own all right. It is noteworthy that when averages are taken the men's wages have been steadily on the increase during recent years. The average pay of all men at the period here concerned was 3·97 shillings per shift; of hewers 4·54. With regard to accidents, nineteen deaths resulted from explosions in 1900, that is 0·1 per thousand men employed; by falls of rock and coal more accidents happened, resulting in 199 deaths, or 1·118 per thousand men, another fruitful source of accident is in the shafts and inclines, not while travelling in the cage but by the trammer running the tubs quickly to the shaft or incline, and being unable to stop the tub, it runs over the edge and draws the man with it; there were 111 fatal accidents in shafts and inclines, or 0·624 per thousand men, and mostly of this character.

The men are well looked after, excellent bathing arrangements and changing rooms, where the home clothes are left and aired while the men are at work and *vice versâ*. In these halls, each man is allotted a place, and is provided with a hook and a pulley high up near the roof, and a length of rope, by this means he can pull the hook up or let it down as desired. When he arrives he lets down the hook on which his pit clothes hang, changes, puts his home clothes on the hook, hoists it up to near the roof, padlocks the end of the rope so that none but he can get the clothes down. There he leaves them until required after the shifts work and the shower bath. Fig. 20 is a view of the changing room at Shamrock Colliery.

There is a general friendly society which embraces most of the collieries and numbers some 120,000 members; in 1899 it helped 23,163 invalids, 15,194 widows and 49,000 orphans. Its receipts were £952,082, its expenditure £728,689, and it has funds, free from debt, amounting to £1,212,741. A good feature about these collieries is the good houses the companies provide in their private villages for the miners, who go home clean, and so do not soil the home and make it unattractive as is too frequently the case in this country. Figs. 21 and 22 are views of the village recently constructed at Schlägel and Eisen Colliery.

The writer wishes, in conclusion, to acknowledge gratefully the uniform courtesy and kindness extended to him by the engineers and functionaries generally at the numerous mines he visited. More particularly, he is pleased to thank the editor of " Glückauf "; General Director Bergrat Behrens, of Bergwerksgesellschaft, Hibernia, Herne; the Harpener Bergbau-Actien Gesellschaft, Dortmund; Haniel and Lueg, Düsseldorf; the Gutehoffnungshütte A.V., Oberhausen, and Aug. Klönne, Dortmund; for photographs, etc., from which many of the figures have been reproduced.

Furthermore, for the benefit of those who may desire further information relating to this coal field, it may be noted that much of the statistical matter included in this article is to be found greatly amplified in the " Festchrift," published on the occasion of the Allgemeiner deutsche Bergmannstag in Dortmund.

## Mr. JOHN GAVEY, C.B., M.Inst.C.E., M.I.E.E.,

### Engineer-in-Chief and Electrician to the Post Office.

IT is safe to assert that there have been few developments in the telegraph and telephone services of the Post Office with which the subject of the present sketch has not been prominently identified. Mr. Gavey was born and educated in Jersey. He entered the service of the Electric and International Telegraph Company in 1860, and when the State assumed control of the telegraphs in 1870 his services were transferred to the Government. He acted successively as superintendent of the South-Eastern, Great Western and South Wales districts. In 1892 came the appointment of Chief Technical Officer at the General Post Office ; in 1897 that of Second Assistant Engineer-in-Chief, in 1899 Assistant Engineer-in-Chief and Electrician, while in 1902 he was called upon to assume the responsibilities attaching to the post of Engineer-in-Chief and Electrician.

Mr. Gavey's career has been one of strenuous application, and he has found time for a great deal of original research Turning to the Electrical Trades Directory we are reminded that when in Bristol he took one of the earliest steps in the direction of increasing the speed of Wheatstone working by designing the arrangement of placing the local apparatus, which indicates when readjustment of the relays of a repeater set is necessary, in a high resistance leak on the main circuit, this leak being so proportioned to the line and apparatus resistance that the current actuating the relay is about the equivalent of that received at the distant station.

A little later, on the initiative of Sir William Preece, he applied himself to the improvement of the repeaters which were placed at Haverfordwest on the Irish wires. At that time, owing to the inertia of the relaying sounders then employed, the insertion of repeaters actually decreased the speed of Wheatstone working as compared with the results obtained in fine weather on direct circuits, and they were only brought into use when the insulation conditions became such that direct working was impossible. Following up the Bristol experiment he introduced repeating direct from the relays, an alteration rendered possible by the use of leak circuits, with the result that the speed of Wheatstone signalling, instead of being reduced, was immediately doubled on the insertion of the repeaters, which thenceforth were constantly retained in circuit.

About the date of his removal to Cardiff the telephone was introduced, and he forthwith devoted his attention to this new development. The result was that by the summer of 1881 the first trunk telephone line connecting

exchanges in two separate post towns in Great Britain (Cardiff and Newport) was erected and brought into use.

Subsequently Mr. Gavey was requested by Sir William Preece to carry out a series of experiments on wireless telegraphy, which have been described in papers read before various institutions. In the course of some of these trials Mr. Gavey obtained the first successful result in so-called " wireless telephony," having in 1899 obtained articulate speech between two parallel wires on opposite sides of Loch Ness. This led subsequently to the establishment of a permanent wireless telephonic service between the rocky islets known as the Skerries and the mainland near Holyhead.

Later, when the Post Office decided on the acquisition of the trunk telephone system in the United Kingdom, he was entrusted with the duty of valuing the trunk lines to be acquired from the National Telephone Company, and subsequently with the organisation and development of that system.

When the extension of the local service of the Post Office was decided upon, he visited various countries in Europe and America to inquire into and report on the systems in actual use, and finally he was called upon, amongst his other duties, to design and supervise the construction of the Post Office system of telephone exchanges in the metropolitan district.

Whilst in Cardiff he took an active share in the educational development of the town. He was for several years a member of the free library and museum, and of the Technical Education Committees. He was for a considerable period hon. secretary, and finally president, of the Cardiff Naturalists' Society. He has amongst other contributions, read the following Papers at the Institution of Electrical Engineers : " On Earth Borers for Telegraph Poles " ; " On Insulators for Aerial Telegraph Lines " ; " On the Telephone Trunk Line System in Great Britain " ; " On Telegraphs and Telephones at the Paris Exhibition " ; and on " Telephones " before the Society of Arts ; in addition to Papers of purely local interest and contributions to the technical press. He acted as juror in the electrical section of the Paris Exhibition of 1900, and was appointed an official delegate to the International Electrical Congress which was held in connection with the Exhibition.

We may add that he attended the Wireless Telegraph Conference at Berlin as one of the British delegates. The C.B. was conferred on him in 1893 in recognition of his public service.

# Admiral Sir John Fisher, G.C.B.

## AN APPRECIATION BY MR. W. T. STEAD.

THE personality of Sir John Fisher has been lately more prominent than ever in connection with the scheme which Lord Esher, Sir George Clarke, and the gallant admiral, have drawn up for the reorganisation of the War Office. We have rarely seen a better appreciation of the admiral than that which is given in the current number of "The Review of Reviews," by Mr. W. T. Stead. It is a curious fact that Admiral Fisher was the last midshipman received into the service by Sir William Parker—the last of Nelson's captains.

Eighteen years ago, says Mr. Stead, when I began my investigations into the state of the navy, I was told by those who knew the service from the top to the bottom that I would find no abler officer afloat or ashore than one Captain Fisher, who had commanded the *Inflexible* at the bombardment of Alexandria, and who was at that moment captain of the *Excellent.*

I sought an introduction to him, which I obtained with some difficulty, for the rules of the service against giving any information to the press were very strict. When I used to go to Captain Fisher, like Nicodemus, at night-time, meeting him at wayside railway stations, I found him, wherever I met him, always the same—one of the pleasantest, frankest and clearest-sighted of men. "Fisher," said an admiral to me in those days, "is the one man we have got who can be compared to Nelson. If Britain were involved in a great naval war, Fisher could achieve as great renown as that of Lord Nelson." His subsequent career has fully justified the confidence expressed in him by his superior officers.

Admiral Fisher since then has commanded the Mediterranean fleet, and it is no exaggeration to say that it is largely owing to the splendid state of efficiency of that fleet under his command that the peace of Europe was maintained in the critical years when the whole land fighting force of the Empire was absorbed in South Africa. He is a supreme type of the modern naval officer at his best. Although sixty-two years of age, Admiral Fisher is in the full vigour of manhood, and as hearty a boy as he was in the days when he first joined the navy in the Crimean War. When he represented the navy at the conference at The Hague, where he did admirable service, he was known as "the dancing admiral." And even now, when commander-in-chief at Portsmouth dockyard, he still thinks nothing of attending ten dancing parties in a fortnight takes part in every dance. and does not go home till three o'clock in the morning. He is brimful of vigour, energy and buoyant vitality. But for all his devotion to the dance, no man is a keener student, nor has any one a more masterly grasp of all the latest improvements in naval warfare.

He is a man born to command, who inspires confidence alike in his superiors and among his subordinates, Nelson, as may well be imagined, is the god of his idolatry. He is saturated in every fibre with the Nelsonian tradition. He has served his country on almost every naval station, he has been a sea lord at the Admiralty, and sooner or later will take his proper place as the first sea lord at Whitehall. On listening to his brilliant conversation, every sentence of which is double-shotted with wit and common sense, I have been constantly reminded of two men who, however diverse from each other and from him, nevertheless possess one great characteristic in common. Admiral Fisher, like Cecil Rhodes and General Gordon, is passionately devoted to his country, and, like them, is vehemently impatient of all the mediocrities who, shackled in red tape, exhaust all their energy in the mere detail of administration, and have neither time nor capacity left for attending to the proper work of direction. Admiral Fisher is a holy terror to skulkers and shufflers, but he has an infinite faith in the capacity of education and discipline. "Give me a boy young enough," he declared, "and I can make anything out of him." For there is in him, as in all great leaders of men, an infinite faith in the latent potentiality of human nature. He is a born optimist, and contact with him kindles enthusiasm even among the dullards. Few men have so great a gift of forcible expression ; his conversation teems with apothegms. But there is a jovial geniality about him which makes every one feel at his ease. If so be that it is necessary to call in the aid of a sailorman in order to advise as to the best method of reforming the administration of the War Office, no better choice could have been made than that of Admiral Fisher.

He enjoys to an almost unprecedented extent the confidence of his King and of his country, while as for the navy, there would probably be a unanimous vote in the service if all sailormen ashore and afloat were to be asked to vote as to what great sea captain of our time was best qualified to lead the navy of Great Britain to victory in a great naval war.

*From the " Review of Reviews."*]

ADMIRAL SIR JOHN ARBUTHNOT FISHER, G.C.B.

(Commander-in-Chief at Portsmouth),

Who, with Lord Esher and Sir George Clarke, drew up the reorganisation scheme for the War Office.

The Admiral lately received the honour of a private visit from His Majesty the King, who inspected Portsmouth Dockyard, etc., and on board the *Alberta, en route* for Cowes, witnessed some special evolutions carried out by submarines and destroyers.

# NOTES AND NEWS.

### The Power Station at Brattforsen.

THE accompanying illustration shows a notable Swedish power station. This, says Affarsvarlden, has been extended so that it can now supply as much as 2,500 electrical h.p.

The material used for the Brattforsen dam is hewn granite, and it rests upon the solid rock; its pillars and front course are embedded in cement, and including the abutments, it has a total length of about 150 metres and an average height of 8·5 metres. On the north bank of the river, about 20 metres below the dam, the power-station has been built. The water is conducted to the spot by two tubes, with diameters respectively of 3 and 2 metres. The height of the fall of water is from 13 to 15 metres, varying with the backwater.

There is nothing to prevent the Orebro Electrical Company supplying considerable additional quantities of power, for arrangements have been made at Brattforsen for utilising the waterfall to the whole of its height, e g., 22 metres, by building an extension of the dam and carrying out canalisation, etc., to some extent. The principal part of the machinery is, moreover, arranged so as to facilitate the utilisation of the fall in its entire height.

The British Westinghouse Electric and Manufacturing Company, Ltd., announce that they are now in a position to receive visitors at their new office at Market Place Buildings, Sheffield, which has hitherto been used only for correspondence. This new branch office is in charge of Mr. C. A. Slater, and it is anticipated that it will prove a great convenience to clients.

Since January 1st, the business of the Electrical Transmission Company, Albert Works, Hammersmith, has been amalgamated with that of the Sturtevant Engineering Company, Ltd., whose works are at 29, Bankside, S.E., ; offices, 147, Queen Victoria-Street, E.C. By this step the very extensive range of all forms of motor starting and controlling devices manufactured by the Sturtevant Engineering Company Ltd., has been augmented by the addition of the specialties in switchgear manufactured by the Electrical Transmission Company.

The Bankside works will be devoted to the manufacture of all apparatus connected with the higher forms of motor control, such as is used for lifts, winches, hoists, haulage, cranes, pumps, etc., and every description of automatic controlling devices. The Hammersmith works will be devoted to the manufacture of the E.T.C. specialties, such as main switches, traction accessories, feeder and section pillars for power and lighting, service switches, motor starters, and speed regulators, etc. The whole field of motor control and switch work is thus covered by the standard apparatus of this combination.

The commercial management of both concerns will be directed from the head office of the Sturtevant Engineering Company, Ltd., at 147, Queen Victoria Street, E.C., to which all correspondence should be addressed.

The battleship *New Zealand*, launched at Portsmouth, sister ship to *King Edward VII.*, is the largest warship ever built at Portsmouth Dockyard.

AT BRATTFORSEN, SWEDEN.

Messrs. Graham Morton and Co., Ltd., of Leeds and London, have just secured contracts for the building of eleven bridges for the Great Western Railway Company and fifty-seven girder bridges for the Bengal North-Western Railway Company.

We understand that Messrs. Richardsons, Westgarth, and Co., Ltd., of Hartlepool Engine Works, have laid down a special plant for the manufacture of steam turbines under license from the Hon. C. A. Parsons.

Mr. A. E. Aspinall is leaving the service of the British Westinghouse Electric and Manufacturing Company, Ltd., for that of the De Beers Consolidated Mines, Ltd., Kimberley, and will act in the capacity of manager of the Costs Department.

A fine display of coal-cutting machinery and electrical plant, coal conveyors, rock drills, pumping machinery, etc., is promised at the second Colliery Exhibition, which will be held at the Royal Agricultural Hall from June 25th to July 2nd.

With the approval of the Council of the Sanitary Institute, Mr. Scott-Moncrieff's sewage-testing apparatus can now be seen in the Parkes Museum. The apparatus has been designed for the purpose of obtaining exact information upon which to base bacterial sewage disposal schemes, particularly as to (1) the depth of filter required to produce the necessary standard of purity in the effluent ; (2) the quantity of air necessary for the life processes of the organisms in the filter ; (3) the correct rate of flow per unit of filter-bed surface in order to obtain the best results ; and (4) the best period of rest between each discharge to prevent galatinous growths in the filtering material.

Messrs. the Taff Vale Railway Company have placed an order with Messrs. the Bristol Wagon and Carriage Works Company, Ltd., of Bristol, and Messrs. the Avonside Engine Company, of Bristol, for six steam motor cars for passenger traffic to run on their railway. These are duplicates of the one designed by the Railway Company's Locomotive Superintendent, Mr. T. Hurry Riches, which has been working so successfully during the past few months. The Avonside Engine Company are building and supplying the locomotive portion to the Bristol Wagon and Carriage Works Company, who are making the carriage portion. The cars are to be delivered for the summer traffic.

At the invitation of the Main Drainage Committee of the London County Council, a number of ladies and gentlemen were present at the formal opening by Lord Monkswell, the Chairman of the Council, of the new pumping station just completed at Lots Road, Chelsea. Lord Monkswell mentioned that the pumping stations of the Council could now deal with over 600,000 gallons a minute. The total cost of the new station has been about £82,000, of which the buildings (erected by the Works Department) accounted for about £50,000, the pumps by Messrs. Easton and Co. for about £6,100, and the engines by Messrs. Crossley Brothers for about £10,300.

The eleventh annual conversazione of the Junior Engineering Society proved a most enjoyable function, an exhibition of mechanical appliances including Mr. Churchward's model of the G.W.R. valve gear, and a locomotive model built by Mr. M. R. Clarke. A two-cylinder, cross-compound, oil-fired locomotive model by Mr. J. C. Crebbin; together with that gentleman's four-cylinder tandem-compound locomotive model, came in for much investigation, and there were many other interesting features. The Society medal was presented to Mr. D. G. Slatter for his paper, "Instruction in Modern Gas-Holders."

A LARGE PELTON WHEEL IN USE NEAR PORT TALBOT.

### Large Pelton Wheel Installation.

This illustration shows one of two large Pelton wheels installed by Messrs. Gilbert Gilkes and Co., Ltd., at the Tinplate Works of the Copper Miners' Tinplate Company, Ltd., Cwm Avon, near Port Talbot, South Wales. These wheels are employed for driving 19-in. rolls used for rolling out thin sheets.

### Screw for 150-ton Sheer Legs.

We illustrate, by the courtesy of Messrs. W. Somers and Co., Ltd. of Halesowen, near Birmingham, an enormous screw, designed for 150-ton sheer legs which are being erected at Chatham Dockyard. The screw will be used for moving the back leg of the sheers. It weighs 17½ tons and was forged from a single ingot of steel. It is 85 ft. 7 in. long, and 11½ in. in diameter, and has a 2 in. thread.

17½-TON SCREW FOR SHEER LEGS
Forged by Messrs. W. Somers and Co., Ltd., of Halesowen, near Birmingham, from a single ingot.

# SLIDE RULES
### FOR THE
# MACHINE SHOP.

BY

### CARL G. BARTH.

The following account of slide rules for the machine shop, as part of the Taylor system of management, was given by the author to the American Society of Mechanical Engineers, at their New York meeting in December. The author has been engaged in making a series of experiments in order to establish data in regard to resistances in cutting steel with edged tools, consequent upon the introduction of high speed steel, and he describes the most interesting of slide rules used in connection therewith.—ED.

BY the use of the slide rules described in the following pages, a most complex mathematical problem may be solved in less than a minute. The author will confine his attention to slide rules for lathe, and will take for an example an old-style belt-driven lathe with cone pulley and back gearing.

Considering the number of variables that enter into the problem of determining the most economical way in which to remove a required amount of stock from a piece of lathe work, they may, be enumerated as follows :—

1. The size and shape of the tools to be used.
2. The use or not of a cooling agent on the tool.
3. The number of tools to be used at the same time.
4. The length of time the tools are required to stand up to the work (life of tool).
5. The hardness of the material to be turned (class number)
6. The diameter of this material or work.
7. The depth of the cut to be taken.
8. The feed to be used.
9. The cutting speed.
10. The cutting pressure on the tool.
11. The speed combination to be used to give at the same time the proper cutting speed and the pressure required to take the cut.
12. The stiffness of the work.

All of these variables, except the last one, are incorporated in the slide rule, which, when the work is stiff enough to permit of any cut being taken that is within both the pulling power of the lathe and the strength of the tool, may be manipulated by a person who has not the slightest practical judgment to bear on the matter ; but which as yet, whenever the work is not stiff enough to permit of this, does require to be handled by a person of a good deal of practical experience and judgment.

However, we expect some day to accumulate enough data in regard to the relations between the stiffness of the work and the cuts and speeds that will not produce detrimental chatter, to do without personal judgment in this matter also, and we will at present take no notice of the twelfth one of the above variables, but confine ourselves to a consideration of the first eleven only.

Of these eleven, all except the third and tenth enter into relations with each other that depend only on the cutting properties of the tools, while all except the

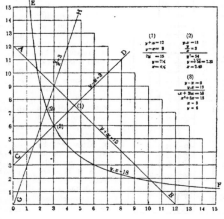

FIG. I.

second, fourth, and ninth also enter into another set of relations that depends on the pulling power of the lathe, and the problem primarily solved by the slide rule is the determination of that speed combination which will at the same time most nearly utilise all the pulling power of the lathe on the one hand, and the full cutting efficiency of the tools used on the other hand, when in any particular case under consideration values have been assigned to all the other nine variables.

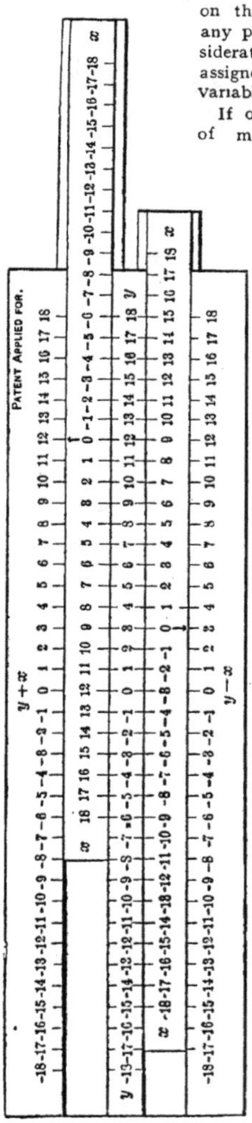

If our lathe were capable of making any number of revolutions per minute between certain limits, and the possible torque corresponding to this number of revolutions could be algebraically expressed in terms of such revolutions, then the problem might possibly be reduced to a solution, by ordinary algebraic methods, of two simultaneous equations containing two unknown quantities; but as yet no such driving mechanism has been invented, or is ever likely to be invented, so that, while the problem is always essentially the solution of two simultaneous equations, or sets of relations between a number of variables, its solution becomes necessarily a tentative one; or, in other words, one of trial and error, and involving an endless amount of labour, if attempted by ordinary mathematical methods; while it is a perfectly direct and remarkable simple one when performed on the slide rule.

FIG. 2.

## THE SLIDE RULE FOR SIMPLE PROBLEMS.

The slide-rule method of solution may, however, also be employed for the solution of numerous similar problems that are capable of a direct and perfect algebraic solution; and it will, in fact, be best first to exhibit the same in connection with the simplest imaginable problem of this kind.

In the first place, the solution of two simultaneous equations may be graphically effected by representing each of them by a curve whose co-ordinates represent possible values of the two unknown quantities or variables, for then the co-ordinates of the point of intersection of these curves will represent values of the unknown quantities that satisfy both equations at the same time.

*Example* 1.—Thus, if we have $y + y = 12$ and $y - x = 3$, these equations are respectively represented by the two straight lines $AB$ and $CD$ in fig. 1; and as these intersect at a point (1) whose co-ordinates are $x = 4\frac{1}{2}$ and $y = 7\frac{1}{2}$, these values will satisfy both equations at the same time.

*Example* 2. Suppose, again, that we have $xy = 18$ and $\frac{y}{x} = 3$, and these equations are respectively represented by the equilateral hyperbola $EF$ and the straight line $GH$; and the co-ordinates of the point of intersection of these (2) being respectively $x = 2.45$ and $y = 7.35$, these values will satisfy both equations at the same time.

*Example* 3. Similarly, if we have $y - x = 3$ and $y.x = 18$, these equations are respectively represented by the straight lines $CD$ and the equilateral hyperbola $EF$; and the co-ordinates to the point of intersection of these (3) being $x = 3$ and $y = 6$, these values will satisfy both equations at the same time.

The slide rule method of effecting these solutions —to the consideration of which we will now pass— will readily be seen to be very similar in its essential nature to this graphical method, though quite different in form.

In fig. 2 is shown a slide rule by means of which may be solved any problem within the range of the rule of the general form : "*The sum and difference of two numbers being given, what are the numbers ?* "

The rule is set for the solution of the case in which the sum of the numbers is 12 and their difference 3, so that we may write

$$y + x = 12 \text{ and } y - x = 3,$$

which are the same as the equations in Example 1 above.

In the rule the upper fixed scale represents possible values of the sum of the two numbers to be found, for which the example under consideration gives $y + x = 12$, opposite which number is therefore placed the arrow on the upper slide.

The scale on this slide represents possible values of the lesser of the two numbers (designated by $x$) and the double scale on the middle fixed portion of the rule represents possible values of the greater of the two numbers (designated by $y$); and these various scales are so laid out relatively to each other, and

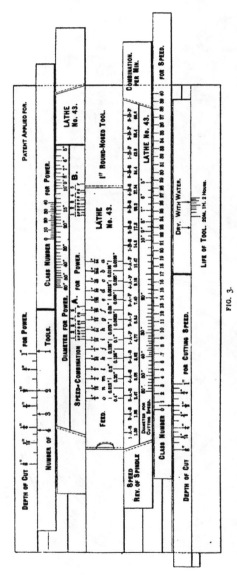

FIG. 3.

to the arrow referred to, that any two coincident numbers on these latter scales have for their sum the number to which this arrow is set; in this case accordingly 12.

The bottom fixed scale on the rule represents possible values of the difference of the two numbers, in this case 3, opposite which number is therefore placed the arrow on the bottom slide of the rule, the scale on which also represents possible values of the lesser of the two numbers, $x$; and the double fixed scale in the middle of the rule representing, as already pointed out, possible values of $y$, the whole is so laid out that any two coincident numbers on these latter scales have for their difference the number to which this arrow is set; in this case accordingly, 3.

Fixing now our attention on any number on the double $y$ scale in the middle of the rule, we first note the values coincident to it in the two $x$ scales on the slides; and, this done, we readily discover in which direction we must move along the first scale in order to pick out that value of $y$ which has the same value of $x$ coincident with it in both $x$ scales. For the case under consideration this value of $y$ is $7\frac{1}{2}$, and the coincident value in both scales is $4\frac{1}{2}$. Evidently, therefore, $y = 7\frac{1}{2}$ and $x = 4\frac{1}{2}$ are the numbers sought.

In the same manner we make a slide rule for the solution of the general problem: "*The product and quotient of two numbers being given, what are the numbers?*"

Such a rule would differ from the above described rule merely in having logarithmic scales instead of plain arithmetic scales.

By the combined use of both arithmetical and logarithmetic scales we may even construct rules for a similar solution of the general problems: "*The sum and product, or the sum and quotient, or the difference and product, or the difference and quotient, of two numbers being given, what are the numbers?*" and a multiplicity of others; and the writer ventures to suggest that slide rules of this kind, and some even simpler ones, might be made excellent use of in teaching the first elements of algebra, as they would offer splendid opportunities for illustrating the rules for the operations with negative numbers, which are such a stumbling block to the average young student.

We now have sufficient idea of the mathematical principles involved, for a complete understanding of the working of the slide rule whose representation forms the main purpose of this article.

### AN IDEAL FORM OF SLIDE RULE.

This slide rule, in a somewhat ideal form in so far as it is made out for neither steel nor cast iron, but for an ideal metal of properties between these two, is illustrated in fig. 3. It will be seen to have two slides in its *upper section* and three in its *lower section*, and is in so far identical with the

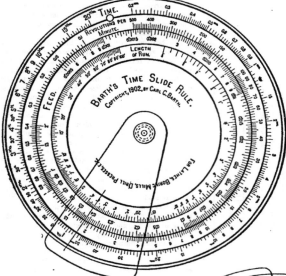

FIG, 4

on the middle cone step, the back gear in, and the slow speed of the countershaft ; and similarly, the combination 1—*B*—*F* designates the belt on the largest cone step on the machine, the back gear out, and the fast speed of the countershaft and so on.

The double, fixed scale in the middle of the rule (marked FEED) is equivalent to the *y* scale of the rule in fig. 2, and the scales nearest to this on the slides on each side of it (marked SPEED COMBINATION FOR POWER, and FOR SPEED, respectively) are equivalent to the *x* scales on the rule in fig. 2. The rest of the scales represent the various other variables that enter into the problem of determining the proper feed and speed combination to be used, fixed values being either directly given or assigned to these other variables, in any particular case under consideration.

rules made for the Bethlehem Steel Company, while in the rules more recently made it had been found possible and convenient to construct it with only two slides in the lower section also.

It is shown arranged for a belt-driven lathe (No. 43*) with five cone steps, which are designated respectively by the numbers 1, 2, 3, 4, 5, from the largest to the smallest on the machine. This lathe has a back gear only, and the back gear in use is designated by the letter *A*, the back gear out by the letter *B*. It also has two countershaft speeds, designated respectively by *S* and *F*, such that *S* stands for the slower, *F* for the faster of these speeds.

The Speed Combination 3—*A*—*S* thus designates—to choose an example—the belt

* The main frame of the rule is used for a number of lathes, and is arranged to receive interchangeable specific scales for any lathe wanted, as may be seen in the illustration.

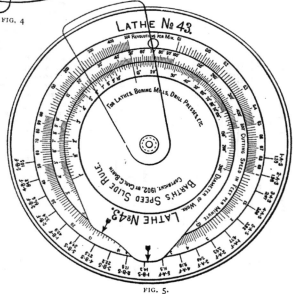

FIG. 5.

The upper section of the rule embodies all the variables that enter into the question of available *cutting pressure* at the tool, while the lower section embodies all the variables that enter into the question of *cutting speed ;* or, in other words, the upper section deals with the *pulling power* of the lathe, the lower section with the *cutting properties* of the tool ; and our aim is primarily to utilise, in every case, both of these to the fullest extent possible.

### SETTING RULE.

The example for which the rule has been set in the illustration is .

A $\frac{1}{2}$-in. depth of cut to be taken with each of two tools on a material of class 14 for hardness, and of 20 in diameter, and the tools to last 1 *hour* and 45 *minutes* under a good stream of water.

The steps taken in setting the rule were :

1. The first scale in the upper or POWER section of the rule, from above, was first set so that 2 in the scale marked NUMBER OF TOOLS became coincident with $\frac{1}{2}$ *in.* in the fixed scale marked DEPTH OF CUT FOR POWER

2. The second slide in this section of the rule was so set that 20 in. in the scale marked DIAMETER OF WORK FOR POWER became coincident with 14 in the scale marked CLASS NUMBER FOR POWER.

3. The first slide from below, in the lower or SPEED section of the rule was so set that the arrow marked WITH WATER became coincident with 1 *hour* 45 *minutes* in the fixed scale marked LIFE OF TOOL.

4. The arrow on the lower side of the second slide in this section of the rule was set to coincide with $\frac{1}{4}$ *in.* in the scale marked DEPTH OF CUT FOR CUTTING SPEED.

5. The third and last slide in this section was so set that 20 in. in the scale marked DIAMETER OF WORK FOR CUTTING SPEED became coincident with 14 in the scale marked CLASS NUMBER FOR CUTTING SPEED.

Let us now separately direct our attention to each of the two sections of the rule.

### POWER SECTION.

In the POWER section we find that all the speed combinations marked $B$ (back gear out), entirely beyond the scale of feeds, which means that the estimated effective pull of the cone belt reduced down to the diameter of the work, does not represent enough available cutting pressure at each of the tools to enable a depth of cut of $\frac{1}{2}$ in. to be taken with even the finest feed of the lathe. Turning, however, to the speed combinations marked $A$ (back gear in), we find that with the least powerful of them ($5-A-F$) the $e$ feed, which amounts to $\frac{1}{25}$ inch — $0\cdot039$ in., may be taken ; while the $f$ feed, which amounts to $\frac{1}{20}$ in. = $0\cdot05$ in., is a little too much for it, though it is within the power of the next combination ($5-A-S$), and so on until we finally find that the most powerful combination ($1-A-S$) is nearly capable of pulling the $i$ feed, which amounts to $\frac{1}{10}$ in. = $0\cdot1$ in.

### SPEED SECTION.

In the SPEED section of the rule we likewise find that all the $B$ combinations lie beyond the scale of feeds, while we find that the combination $5-A-F$ (which corresponds to a spindle speed of $11\cdot47$ revolutions per minute), can be used in connection with the finest feed ($a$) only, if we are to live up to the requirements set for the life of the tool ; while the next combination ($4-A-F$) will allow of the $e$ feed being taken, the combination $3-A-F$ of the $f$ feed, and so on until we finally find that the combination $3A$ is but a little too fast for the coarsest ($o$) feed, and that both of the slowest combinations ($1-A-S$ and $2-A$ is $-S$) would permit of even coarser feeds being taken, so far as only the lasting qualities of the tools are concerned.

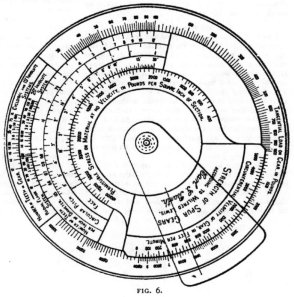

FIG. 6.

We thus see that there is a vast difference between what the Power section of the rule gives as possible combinations of feeds and speeds for the utilisation of the full pulling power of the lathe, and what the Speed sections of the rule gives for such combinations for the utilisation of the tools up to the full limit set., However, by again running down the scale of feeds we find that, in both sections of the rule, the 1 feed $\frac{1}{16}$ in. = 0·1 in.), is but a trifle too coarse for the combination 1—A—F, while the h feed ($\frac{3}{84}$ in. = 0 078 in.) is somewhat too fine in connection with this speed combination 1—A—F, both for the full utilisation of the pulling power of the belt on the one hand, and for the full utilisation of the cutting efficiency of the tools on the other hand.

In this case, accordingly, the rule does not leave a shadow of doubt as to which speed combination should be used, while it leaves us to choose between two feeds, the finer of which does not allow us to work up to the full limit of either the belt or the tools, and the coarser of which will both overload the belt a trifle and ruin the tools a trifle sooner than we first intended to have them give out.

The final choice becomes a question of Judgment on the part of the *Slide Rule and Instruction Card Man,* and will depend upon how sure he is of having assigned the correct Class Number to the material or not ; and this latter consideration opens up a number of questions in regard to the practical utilisation of the rule, which for the lack of time cannot be taken up in the body of this article.

#### THE TIME SLIDE RULE.

Having decided upon the speed and feed to use, the Instruction Card Man now turns to the Time slide rule illustrated in fig. 4. and by means of this determines the time it will take the tools to traverse the work to the extent wanted, and making a fair allowance for the additional time consumed in setting the tools and calipering the work, he puts this down on the instruction card as the time the operation should take.

For finishing work the pulling power cuts no figure, so that this resolves itself into a question of feed and speed only ; and for the selection of the speed combination that on any particular lathe will give the nearest to a desired cutting speed, the Speed slide rule illustrated in fig. 5 is used.

It will readily be realised that a great deal of preliminary work has to be done before a lathe or other machine tool can be successfully put on a slide rule of the kind described above. The feeds and speeds and pulling power must be studied and tabulated for handy reference, and the driving belts must not

be allowed to fall below a certain tension, and must in every way, be kept in first-class condition.

In some cases it also becomes necessary to limit the work to be done, not by the pull that the belts can be counted on to exert, but by the strength of the gears, and in order to quickly figure this matter over the writer also designed the Gear slide rule illustrated in fig. 6, which is an incorporation of the formulæ established several years ago by Mr. Wilfred Lewis.

For the pulling power of a belt at different speeds, the writer has established two formulæ, which take account of the increasing sum of the tensions in the two sides of a belt with increasing effective pull, and which at the same time are based on the tensions recommended by Mr. Taylor in his paper entitled " Notes on Belting." These formulæ have also been incorporated on a slide rule.

#### CONCLUSIONS.

Having thus given an outline of the use of the slide-rule system of predetermining the feeds and speeds, etc., at which a machine tool ought to be run to do a piece of work in the shortest possible time, the writer, who has made this matter an almost exclusive study during the last four years, and who is at present engaged in introducing the Instruction Card and Functional Foremanship System into two well-known Philadelphia machine shops, which do a great variety of work in both steel and cast iron, will merely. add that, in view of the results he has already obtained, in connection with the results obtained at Bethlehem, the usual way of running a machine shop appears little less than absurd.

Thus already during the first three weeks of the application of the slide rules to two lathes, the one a 27 in. the other a 24-in., in the larger of these shops, the output of these was increased to such an extent that they quite unexpectedly ran out of work on two different occasions, the consequence being that the superintendent, who had previously worried a good deal about how to get the great amount of work on hand for these lathes out of the way, suddenly found himself confronted with a real difficulty in keeping them supplied with work. But while the truth of this statement may appear quite incredible to a great many persons, to the writer himself, familiar and impressed as he has become with the great intricacy involved in the problem of determining the most economical way of running a machine tool, the application of a rigid mathematical solution to this problem as against the leaving it to the so-called practical judgment and experience of the operator, cannot otherwise result than in the exposure of the perfect folly of the latter method.

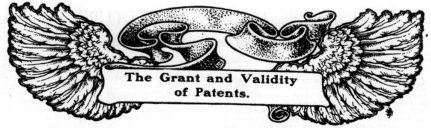

# The Grant and Validity of Patents.

## A NEW WORK OF REFERENCE FOR LAWYERS AND PATENTEES.

MR. JAMES ROBERTS, M.A., LL B., in the course of a new work on "The Grant and Validity of British Patents for Inventions,"* draws attention to a remarkable fact. It has been ascertained by an official enquiry that of the patents for inventions granted in England, no fewer than forty-two per cent., that is about 5,780 per annum, are invalid on the ground of having been already patented in this country. An examination of the results of litigation shows, that of such patents as are commercially worth infringing, no less than fifty-one per cent. are invalid. The invalidity of these patents is, in many cases, not discovered till after a lapse of years, but even assuming that no invalid patent is renewed, sums amounting to £23,120 per annum are paid for patents which give no legal protection to the patentees. This state of affairs, says Mr. Roberts, brings discredit upon all British patents and diminishes the market value even of those which are valid. Under these circumstances we imagine that a work which has been undertaken to enable the inventor to draw up his specification with the best possible chances of passing successfully between the Scylla of the patent office and the Charybdis of anticipated claims, should be welcome to all manufacturers who are concerned in placing new machinery on the market.

The first part consists of the principles and rules affecting the grant and validity of patents, and the practice respecting the amendment of specifications, both before the Comptroller-General and Law Officers of the Crown ; the second, of abstracts of cases illustrating the applications of the principles ; and the third of the statutes and rules.

In dealing with the question of the grant and validity of patents for inventions from the point of view of the inventor, it must be borne in mind, says the author, that, according to the English law and practice, the question of validity (save in a few cases) cannot be entertained or decided by the authorities whose duty it is to grant such patents. The inventor takes his patent at his own risk, and the validity of the grant may be contested in subsequent proceedings in the High Court, either by a petition for revocation being presented, or by the defence in an action for infringement. A "patentable invention" and a "valid claim " mean, therefore, not merely those for which a grant may be obtained, but those which will be upheld and supported in subsequent litigation.

The rules on which the questions affecting the grant and validity of patents depend are derived from various sources : (1) The common law and considerations of public policy ; (2) the Statute of Monopolies, 1624 ; (3) a long series of decisions elucidating the foregoing ; (4) the Patents, etc., Acts, 1883 to 1902, and cases thereon ; and (5) the rules made under the provisions of those Acts. Although the rules relating to validity are mainly found in actions for infringement and petitions for revocation of patents, yet a knowledge of them is necessary in order to avoid taking out a patent which cannot be subsequently maintained when challenged.

Under the new practice introduced by the Act of 1902, the applicant will be informed by the Comptroller of such previous specifications as appear to anticipate the invention in respect of which a patent is applied for. He must, therefore, with professional assistance in most cases, decide whether he will modify his application, and if so, in what manner. This decision cannot be arrived at without a knowledge of the principles upon which the Courts decide on the validity of patents, and the mode of application of those principles.

The rules and their application are dealt with in the new work under *four* main heads or divisions :—

(1) The consideration of the "manufacture" or "invention" for which a patent may be granted, distinguishing it on the one hand from the principle, involved, the application of which constitutes the "invention," and on the other from the resulting advantages and uses to which it may be put, that is to say, where the monopoly begins and ends in relation to the *manufacture.*

(2) The relations arising from the development of the knowledge of the art in question, and the consideration of the "invention " of the manufacture in regard to *time.* On the one hand, there are the questions of novelty, prior user, prior grant to a rival inventor, and the question of the extent to which the proposed grant might interfere with workmen at the time by reason of the slight amount of ingenuity required to produce the invention in question. On the other hand, there is the relation of the patentee to *subsequent* inventors involving the question of how far the inventor can anticipate subsequent inventors, by including that which he had not actually devised at the date of his application.

(3) The persons to whom and the conditions on which the grant will be made. Under this head come the filing of specifications disclosing the method of performing the invention and making distinct claim or claims thereto, and also questions arising from the policy of the law and the rules of construction or interpretation of specifications.

(4) The procedure to be followed, the drafting of specifications, the amendment of specifications, and opposition to the final sealing of the patent.

---

* "The Grant and Validity of British Patents for Inventions." By James Roberts, M.A., LL.B. John Murray. 25s net.

# MODERN STEEL MANUFACTURE.

## A COMPARISON OF THE BESSEMER AND OPEN-HEARTH PROCESSES.

SOME of the most interesting pages in Professor F. W. Harbord's new work on "The Metallurgy of Steel,"* are devoted to a comparison of the Bessemer and open-hearth processes.

The first consideration is the question of quality, and here, says Professor Harbord, it must be admitted that acid processes have the advantage over basic processes working under ordinary conditions, as, assuming care is taken to select suitable pig-iron, the risk of producing high phosphorus steel in the acid processes is at once eliminated, whereas, however great the care taken, the removal of the phosphorus in the basic processes depends upon the skill and attention of the men, and there is always the possibility that the metal may not be dephosphorised to the required extent. In the case, too, of the manufacturer of rails, or other fairly high carbon steel in the converter, there is the additional risk of rephosphorisation from the slag on adding the recarburising material, even when every precaution is taken. With dead soft steel this danger exists in a smaller degree. In respect to removal of other impurities, the acid processes offer no advantages over the basic, as, in the removal of sulphur, the advantage lies with the basic process, in which removal, if somewhat erratic, does frequently take place.

With regard to the open-hearth process where steels of different grades are required, with carbon varying from ·100 to ·70 or 1·00, the acid process has again the undoubted advantage over the basic open hearth, as in working ordinary phosphoric pig-iron, recarburisation would have to be effected outside the furnace by some recarburising process, which does not enable the same control to be maintained as when it can be done in the hearth of the furnace. In the special case of working hematite iron in a basic furnace, the conditions are different, and we have the advantage of the pure iron used in acid furnaces, with the additional advantage that even the small percentages of phosphorus originally present would be greatly reduced, and, consequently, an exceptionally pure material can be made.

Comparing the quality produced in the open hearth, whether acid or basic, with that of the Bessemer processes, there can be no doubt that the former produces a more regular and reliable material. In the first place, the operations are under greater control. Samples can be taken at repeated intervals, and examined if necessary, both chemically and mechanically, and the working of the process modified to produce the desired result. There is also far less danger of over-oxidation. That the quality of open-hearth steel is, for the same class, superior to Bessemer, is confirmed by the fact that engineers, as the result of experience, always prefer it, and are prepared to pay a higher price for it. For high carbon steels which have to conform to rigid specifications, the acid Siemens is almost exclusively employed, as experience has shown that it is extremely difficult to get the exact carbon percentage specified by the

other processes. Exception to this may be made in a few cases of small Bessemer plants in Sweden and this country, working under special conditions as to the regularity and quality of their pig supply, and the smallness of their charges, but, speaking generally, the statement is correct.

### Yield of Metal.

Here the open-hearth processes have a considerable advantage over the Bessemer. The average yield of an acid Bessemer plant making rail steel will vary from about 86 to 88 tons of steel ingots for every 100 tons of metal (including pig, scrap, spiegel, etc.) used; and a basic Bessemer will yield only about 83 tons of ingots, whereas the acid open hearth will yield about 97 and the basic about 95·0 tons of ingot per too tons of metal charged into the furnace. One reason why the yield in the basic processes is lower than the acid, is that the chemical loss due to presence of phosphorus and more manganese is greater.

### Cost of the Acid and Basic Processes.

Both the labour costs and the costs as regards renewals of linings, repairs and general refractories must always be higher in the basic than in the acid processes; but against these disadvantages must be put the lower cost of pig-iron and scrap and the value of the basic slag produced. For the same size of plant, on the other hand, the output of the acid processes will be greater.

### Comparison as to Cost.

Under this heading it is shown that the difference in cost of production by the acid and basic Bessemer is so little, that it is almost impossible to say which has the advantage, although it is probable that the basic is slightly the cheaper.

The acid Siemens costs of production are probably a little higher than any of the others, although they are only slightly so, and, provided a good supply of scrap equal to 30 per cent. of the charges can be obtained, the difference is very slight.

### Future for Basic Open-hearth Steel.

"It seems probable," says the author, "that in the future the basic open hearth, or some modification of it, will gradually supersede both the basic and acid Bessemer, and to some extent the acid open hearth, both for structural and rail steel, as there are immense deposits of ores which are just too phosphoric for acid work, and which would make an ideal pig-iron for basic open-hearth practice. From many of these ores a pig-iron with about ·15 to ·5 per cent. of phosphorus would be produced, and such a pig-iron worked under basic open-hearth conditions would undoubtedly produce a far better material than is on the average produced either by the acid or basic Bessemer, as with such content of phosphorus in the pig-iron the risk of making steel high in phosphorus would be practically nil, or at all events not so great as the risks incurred by using hematite iron dangerously near the limit allowed in acid open-hearth practice.

---

* "The Metallurgy of Steel," by F. W. Harbord, Assoc. R.S.M., F.I.C., with a Section on the Mechanical Treatment of Steel, by J. W. Hall, A.M.Inst.C.E. Charles Griffin & Co., Ltd. 25s. net.

# SHIPBUILDING NOTES.

## Differing Returns.

Since our estimate was made for publication in last number of the output of new ships in 1903 the statistics of Lloyd's Register have appeared. These, as usual, differ to some extent from the shipbuilders' returns upon which we proceed, because Lloyd's do not include vessels under 100 tons unless they are intended for classification, and they account vessels "under construction" so long as they are in builders' hands, whereas we enumerate only the vessels put into the water during the period under review. It is necessary to remember this, and also the manner in which warships and merchant ships are accounted separately, in noting Lloyd's returns. According to these, during 1903, 697 vessels, exclusive of warships, of 1,190,618 tons gross (viz., 632 steamers of 1,165,503 tons and 65 sailing vessels of 25,115 tons) were launched in the United Kingdom. The warships launched at both Government and private yards amounted to 41 of 151,890 tons displacement. The total output of the United Kingdom for the year was, therefore, 738 vessels of 1,342,508 tons. The output of mercantile tonnage in the United Kingdom during 1903 shows a decrease of 237,000 tons from 1902, and is the lowest since 1897. Compared with the returns for 1901, when the output of both mercantile and war tonnage reached the highest level, the figures show a reduction of 334,000 tons as regards merchant vessels, and 60,000 tons as regards war vessels.

## British Output.

Of the total output, 936,776 steam tons and 13,800 sailing tons, or 950,576 tons in all (80 per cent.) belong to ports in the United Kingdom. The losses of United Kingdom vessels during the twelve months are shown by the Register Wreck Returns to average 258,200 tons (202,700 steam, 55,500 sail). The sales to foreign and colonial owners for 1903 reached a total of 352,000 tons (294,000 steam, 58,000 sail). Purchases from foreign and colonial owners during the same period amounted to 65,000 tons (57,000 steam, 8,000 sail). The sailing tonnage of the United Kingdom thus decreased by about 92,000 tons, while the steam tonnage increased by 497,000 tons. The net increase of United Kingdom tonnage during 1903 was, therefore, about 405,000 tons. For the last four years the estimated net increases were as follows :—1899, 313,000 tons ; 1900, 220,000 tons ; 1901, 543,000 tons ; 1902, 643,000 tons. In 1903, 18 per cent. of the total output was acquired by foreign and colonial shipowners, as compared with 18 per cent. in 1902, 23 per cent. in both 1901 and 1900, 19 per cent. in 1899, 22 per cent. in 1898, and 25 per cent in 1897. The British Colonies provided a considerable amount of employment for the shipbuilders of the United Kingdom, viz., 30 vessels of 33,793 tons (2·8 per cent. of the total output). Germany followed with nine vessels of 26,598 tons. Next Norway with 25,813 tons, and Holland with 18,153 tons.

## Large Steamers.

The shipbuilding statistics of recent years have illustrated the tendency towards the construction of steamers of large tonnage. During the four years, 1892-5, on an average eight vessels of 6,000 tons and upwards were launched per annum in the United Kingdom ; and in the following four years, 1896-9, the average rose to 25. For the last four years, 1900-3, the average has been 39. Of vessels of 10,000 tons and upwards, while only three were launched in the four years, 1892-5, 17 were launched during the four years

1896-9 ; and 32 have been launched during the four years 1900-3. The largest steamers launched during 1903 were the following :—

| | Tons gross. | | Tons gross. |
|---|---|---|---|
| Baltic | ... 23,763 | Kenilworth Castle | 13,150 |
| Not named... | ... 16,780 | Armadale Castle... | 12,800 |
| Not named... | ... 16,780 | Macedonia ... | 10,523 |
| Republic .. | ... 15,378 | Marmora .. ... | 10,523 |

The largest sailing vessel launched in the United Kingdom during 1903 was the *Mneme* of 2,456 tons.

## Turbine Steamers.

In view of the increasing employment of the turbine method of propulsion we may note the names of the steamers fitted with turbines launched during 1903. They are as follows :—

| | Tons displacement. | | Tons gross |
|---|---|---|---|
| H.M.S Amethyst.. | 300 | Yacht Lorena ... | 1,303 |
| H.M.S. Eden | ... 565 | Channel Steamers— | |
| | | The Queen ... | 1,676 |
| | | Brighton... ... | 1,129 |

## Foreign Output.

The construction in foreign shipyards during the year, according to Lloyd's, was 549 steamers of 798,205 tons, and 404 sailing vessels of 156,808 tons, in addition to 78 war vessels of 239,210 tons displacement. Among foreign countries, the largest builders were the United States of America (382,000 tons), Germany (184,000 tons), and France (93,000 tons). Of the mercantile tonnage reported from the United States, a considerable proportion does not affect the general commerce of the world, being intended for service on the Great Lakes. Twenty-four steamers were built for this trade during 1903 of upwards of 4,000 tons each, while seven others ranged between 3,000 and 4,000 tons each. On the sea coast, nine steamers of over 4,000 tons each, besides six steel and three wooden sailing vessels of over 2,000 tons each, were launched. The largest steamers included in these figures are the following :—

| | Tons gross. | | Tons gross. |
|---|---|---|---|
| Minnesota | ... 21,000 | Maine ... | ... 7,914 |
| Mongolia | .. 13,638 | Missouri | ... 7,914 |
| Manchuria | .. 13,638 | | |

Germany launched one steamer exceeding 8,000 tons, viz., the *Gneisenau*, of 8,081 tons, built at Stettin, and 13 steamers between 4,000 and 8,000 tons. The only important sailing vessels included in the output of Germany during the year was the four-masted barque *Petschili* of 3,087 tons, built at Hamburg, under the survey of Lloyd's Register. The most significant feature in respect of the shipbuilding industry in France during 1903 was the abandonment of the construction of large sailing vessels. During the years 1899 to 1902 the numbers of steel sailing vessels of 2,000 tons and upwards launched in France were, respectively, 24, 38, 49, and 54. During 1903 not one such vessel was launched. The steam tonnage launched in France during 1903 amounts to 83,000 tons, or 38,000 tons in excess of the output of 1902.

## The World's Output.

The total output of the world during 1903 (exclusive of warships) appears from Lloyd's figures to have been about 2,146,000 tons (1,964,000 steam, 182,000 sail). The Wreck Returns show that the tonnage of all nationalities totally lost, broken up, &c., in the course of the twelve months amounts to about 744,000 tons (419,000 steam, 325,000 sail). The net increase of the world's mercantile tonnage during 1903 would thus be about 1,402,000 tons. Sailing tonnage has been reduced by 143,000 tons, while steam tonnage has increased by 1,545,000 tons. Compared with the net increase for the world the net increase of 405,000 tons for the United Kingdom is equivalent to nearly 29 per cent. In the net increase of the world's steam tonnage, viz., 1,545,000 tons, the United Kingdom shared to the extent of 497,000 tons, or over 32 per cent. Of the tonnage launched during 1903, the United Kingdom acquired over 44 per cent; and of the new steam tonnage the United Kingdom acquired nearly 48 per cent.

### Comparative Totals.

We extract the following from Lloyd's summary in order to compare with the statistics we presented last month:—

*Vessels launched in the United Kingdom during* 1903 :—

|  | Total 1901. | | Total 1902. | |
|---|---|---|---|---|
|  | No. | Tons. | No. | Tons. |
| Merchant and other vessels (not war-ships) ... ... | 697 | 1,190,618* | 694 | 1,427,558* |
| Warships at Royal Dockyards) ... | 4 | 28,290† | 5 | 51,400† |
| Warships at private yards ... ... | 37 | 123,600 | 18 | 42,740 |
| Total ... | 738 | 1,342,508 | 717 | 1,521,698 |

\* Tons gross.          † Tons displacement.

### Allan Turbine Steamers.

With regard to the two new turbine boats which the Allan Company are building, the first of them, the *Victorian*, is to be ready for work in the summer of this year. The turbine machines to Mr. Parsons' designs are being constructed by Messrs. Workman, Clark and Co., Ltd., Belfast, who have secured the right of constructing such engines. The workmanship will be of the highest class, the boiler power ample, and the pumps, valves, condensers, and other allied parts are specially adapted to the designs of Mr. Parsons. A special arrangement has been devised by Mr. Parsons, whereby a reversing power equal to that of the forward propelling power can be imparted to the machinery, securing almost instant arrestment of the ship's forward motion, and speedy backing in case of need. In this matter the *Victorian* is designed to surpass the ordinary steamer. The widespread arrangement of her propellers, of which there are three in number, each on a separate length of shafting, and the rapidity with which power can be directed upon each of the outer shafts separately, in any direction, will greatly assist the ship's manœuvring power. The *Victorian* will be by far the largest steamer and swiftest of the Allan fleet. She will be fitted in the best style for upwards of 1,500 passengers, and is expected, by reason of the absence of vibration inseparable from the ordinary steam engine, and by the rapidity and unbroken steadiness of revolution in her shafting and propellers, to be both noiseless and steady in a seaway, even while steaming at full power. She is expected to save a day in the voyage between Liverpool and Montreal.

### Transatlantic Passenger Traffic.

The return of passengers landed by Transatlantic liners at New York during 1903 shows an increase of about 90,000 of all classes, equal to 12½ per cent., while the number of ship passages is only 4·7 per cent greater. The British lines seem to have done better than the German, the addition to their numbers

being 16½ per cent., against 13½ per cent. by the German vessels; but in cabin passengers the German ships excelled, their increase being about 25 per cent., against about 12½ per cent. for British lines. The German ships carried 37·3 per cent. of all the cabin passengers, as compared with 34 per cent. in the previous year, while the proportion by British ships declined from 37½ per cent. to 34·3 per cent. This is attributed to the high-speed vessels of the Germans. The mails per *Kaiser Wilhelm der Grosse* averaged 152 hours 18 minutes, the *Kronprinz Wilhelm*, 154 hours 18 minutes; the *Kaiser Wilhelm II.*, a third North-German Lloyd liner, 161 hours 6 minutes; and the *Deutschland*, 162 hours 42 minutes, all from New York to St. Martin's-le-Grand. The fifth ship was the the *Lucania*, with 170 hours 36 minutes; next, the *Campania*, 171 hours 18 minutes; the *Oceanic*, 173 hours 6 minutes; the *Philadelphia*, 178 hours 54 minutes; *La Savoie*, 180 hours 12 minutes; and *La Lorraine*, 180 hours 18 minutes. The ships of the Morgan Combine, other than the German liners, decreased their proportion of cabin passengers from 31 per cent. to 28 per cent —their percentage of steerage passengers remaining practically the same, 18·3 per cent.

### Classes of Passengers.

The total number of passengers was 804,795, about double the number landed in 1899, and nearly three times the number of 1894 and 1897, but the number of ships has not increased greatly. Cabin passengers show a greater proportionate advance in numbers than steerage passengers. Thus there were 161,438 cabin passengers in 1903, or 15 per cent. more than in 1902, and nearly double the average of 1894-98. Second-cabin passengers bear a proportion to first-class of 1·38 to 1, the totals being 67,808 first and 93,630 second-class passengers. The steerage passengers numbered 643,358, about 12 per cent. more than in 1902, and the greatest yet recorded. The number of trips last year was 969—the largest number for many years—and the average number of passengers was the highest yet reached, viz., 830, as compared with 773 in 1902, while in 1901 it was 639, and in 1900, 645. Since 1897 it has steadily increased from 313. All the Continental lines increased their totals, and several of these, especially the large emigrant-carrying lines, had a very large average per ship The total passengers in the case of the Holland-American lines work out to a fraction over 1,000, principally in the steerage, the Italian line about 900, the Prince line 874, the Fabre line 816, and the Spanish line 505.

### Cunard and White Star.

The White Star line has the best record for average first-class passengers per ship, and for first and second-cabin average the Cunard yields to the North German Lloyd. The Cunard Company made more trips than in several preceding years, viz., 65, against 51, 57, 51 and 62; and there was thus a reduction in the number of cabin passengers per ship from 320 to 284 in 1902, but an increase in the number in the steerage from 464 to 518. The Cunard had one trip from the Mediterranean—a new service which took 74 second cabin and 241 steerage passengers. The White Star increased the number of passages made, 91 as com pared with 65, 66, 50, and 57. The cabin passengers per trip averaged last year 246, while in 1902 the number was 283, in 1901 275, in 1900 299, and in 1899 223. The average number of steerage passengers was 502 in 1903, while in the previous years it was 619, 462, 587, and 442 respectively.

# LOCOMOTIVE ENGINEERING NOTES.

BY

## CHARLES ROUS-MARTEN.

### A Huge American Locomotive.

America seems determined to keep well in front as regards locomotive magnitude. We, in Great Britain, have gradually and cautiously crept up to a total weight of 73 tons for an express tender engine, exclusive of its tender, as in the case of the newest express engine of the Caledonian and North-Eastern Railways, while the former has a total weight (including her eight-wheel tender) of approximately 130 tons, a new tank engine to which later reference will be made and which has just been constructed for the Lancashire and Yorkshire line, weighs 77½ tons in working order. All of these represent a vast advance on what a comparatively few years ago was deemed an enormous and barely justifiable weight on the rails. Twenty and even fifteen years ago, a tender-engine weighing 50 tons, exclusive of its tender, was something quite unknown in this country. Even ten years ago very few British tender-engines reached, and fewer exceeded, that weight. It was little more than four years ago that Mr. J. A. F. Aspinall brought out what then was regarded as a wonder of bigness, his express engine No. 1400 of the "Atlantic" or 4–4–2– class, which weighed 58½ tons. Later Mr. Wilson Worsdell beat that with 62 and 67 tons respectively, then Mr. W. Dean followed with 69 tons, and now Mr. G. J. Churchward has exceeded 70, while Mr. J. F. M'Intosh's No. 49 of the 4–6–0 type on the Caledonian, and Mr. Worsdell's 4–4–2–, No. 532, etc., on the North-Eastern, are accredited with a weight of 73 tons each, if indeed this be not somewhat exceeded in the case of the Caledonian colossus. And, lastly, we have, as I have said, a Lancashire and Yorkshire tank engine weighing 77½ tons, but that is, of course, engine and tender in one. Yet these enormous dimensions which staggered us at their onset sink into relative insignificance when compared with those of an American locomotive recently brought out by Messrs. Burnham, Williams and Co., at their celebrated Baldwin Works, Philadelphia, U.S.

### A Mighty Tandem Compound Decapod.

When one has used up such picturesque terms as "giant," "colossus," "leviathan," "monster," etc., in depicting our British engines, one is simply unable to devise any descriptive term which will be adequate to indicate the new locomotive. For, in the first place, the engine itself in working order weighs over 127 tons, exclusive of the tender; the tender weighs approximately, 73 tons loaded, so that the total weight of engine and tender at work reaches the almost incredible aggregate of 200 tons, carried on no fewer than twenty-two wheels. I may explain that it has been specially designed and built for heavy goods traffic on the Atcheson, Topeka, and Santa Fé Railway, where the loads are heavy and the gradients severe. The engine may be classed as of the Decapod type, the term being used in its common-sense purport indicating that the machine walks upon ten "feet"; in other words, has ten wheels coupled. The fact of its having a two-wheel pony truck in the rear as well as in front, does not in reality modify its decapod character. Or according to the most recent system of classification, its type is 2–10–2. The coupled wheels are 4 ft. 9 in. in diameter, and the middle pair is driven by four outside cylinders arranged in pairs tandemwise, the two high-pressure cylinders placed in front being 19 in. in diameter, the two low-pressure cylinders immediately

behind them 32 in., while the piston stroke is also 32 in. These ten-coupled wheels have an adhesion weight of over 104 tons. It will thus be obvious that the engine possesses gigantic tractive force, but this stupendous cylinder power would be of little value unless supported by abundant boiler capacity for steam generation. Accordingly, an extended wagon top boiler is supplied, which is no less than 6 ft. 6¼ in. in diameter with a length of 20 ft. between the tube-plates. It has a total heating surface of 4,796 square feet, of which the iron tubes—2¼ in. in diameter—provide 4,586 square feet, and the steel fire-box, which is 9 ft. by 6 ft. 8 in., 210 square feet, the grate area being 58½ square feet. Moreover, it carries the immense steam pressure of 225 lb. to the square inch. Now the mere recital of these amazing dimensions is sufficient to show at a glance to any professional reader what a powerful machine is represented. It will be most interesting to learn of what maximum duty such an engine is capable. In this small country one finds it difficult to realise the length of the goods-trains that can profitably be hauled on such a line as the Santa Fé, which, by-the-by, gives its title, or rather nickname, to the new engine. In Britain there would be no scope for a locomotive so vast; the load of American dimensions is needed to afford full play to its capabilities.

### The Tandem Plan of Compounding.

So far, locomotive engineers all over the world have seemed a little shy of adopting the tandem system of compounding. It was tried on our own Great Western Railway some fourteen or fifteen years ago, two 7 ft. coupled express engines being built on that principle, but for various reasons of detail, into which I need not now enter, the plan did not prove a success, and was accordingly abandoned. Another method of tandem compounding was tried on the North British Railway, being applied to the identical engine that was in the Tay Bridge accident and that had lain for many months at the bottom of the Tay estuary. It is not probable that the long submersion had anything to do with the result of the experiment, but, at all events, the outcome did not give satisfaction, and the engine, No. 224, was rebuilt on the ordinary non-compound lines. The tandem plan has been tried several times on Continental railways, and, indeed, is still in use to a limited extent, but has never hitherto come into very wide favour. One of its most recent and interesting applications abroad has been in the case of some new suburban tank engines designed by Monsieur du Bousquet, Ingénieur-en-Chef du Matériel et de la Traction of the Chemin de Fer du Nord, which have been at work for some months, and, apparently, are giving favourable results. In the case of the new Santa Fé engine the two low-pressure cylinders and their valve chests are supported by a suitable casting secured to the smoke-box, while the high-pressure cylinders with their steam chests are secured to the fronts of the low-pressure cylinders. The valves are of the balanced-piston type with bushings forming the interior of the chests. Connections between the steam chests and the respective high and low-pressure valves are in the form of a slip joint, made tight with a packed gland. The valve is made in two sections, one governing the admission of steam to the high-pressure and the other that to the low-pressure cylinder, but both sections are secured and operated by the same rod. To enable the steam to pass from the front of the high-pressure cylinder to the back of the low-pressure cylinder, and *vice versa* without crossed ports, the steam is exhausted from the high-pressure cylinder by means of the central opening in the valve through the interior of the high-pressure valve and into the body of the steam chest which acts as a receiver. From this it is distributed to the low-pressure cylinder by the action of the low-pressure valve which is set to act in accord with the high-pressure valve and opens the low-pressure ports at the proper time. The final exhaust takes place through the central external cavity of the low-pressure valve which opens a passage to the smoke-box. The tandem method appears well worthy of a further trial in this country, but the time has not yet arrived for judging whether it does in reality possess any superiority to the ordinary method of compounding.

## Britain's Biggest Tank Engine.

Brief reference has already been made to the new tank engine, No. 404, designed by Mr. H. A Hoy, chief mechanical engineer of the Lancashire and Yorkshire Railway, and built at the Horwich Works for the heavy suburban and semi-suburban passenger services of that line—chiefly on the Manchester and Oldham branch, which has gradients as steep as 1 in 41. This massive and splendid-looking engine has a boiler with no less than 2,039 square feet of heating surface—of which 162 square feet are supplied by the Belpaire fire-box—a grate area of 26 square feet, inside cylinders 19 by 26, and six-coupled wheels 5 ft. 8 in. in diameter. Radial axles are given to the leading and trailing pairs of trailing wheels. There are very large tanks and coal bunkers which will hold, respectively, 2,000 gallons of water and 3¾ tons of coal, thus enabling the locomotive to carry as ample provisions in these respects as many fairly large tender-engines. The total weight loaded, namely, 77 tons 10 cwt, is distributed as follows: Leading radial wheels 10 tons 12½ cwt.; front coupled wheels 17 tons 19¼ cwt.; middle coupled wheels 17 tons 4¼ cwt.; hind coupled wheels 17 tons 4 cwt.; trailing radial wheels 14 tons 10¼ cwt. A pick-up water-scoop is provided so as to minimise the number of times that water has to be taken in while the engine is at a stand. The new type is obviously of huge power and should prove exceedingly useful on the Lancashire and Yorkshire Railway with its heavy loads, steep grades, and frequent stoppages.

## The New Great Western Double-ender.

There is some similarity in general design between the engine just referred to and the newest type of tank engine, No. 99, designed and built by Mr. G J. Churchward, locomotive superintendent, for the Great Western Railway. Like the Lancashire and Yorkshire engine, it has six-coupled wheels 5 ft. 8 in. in diameter with leading and trailing carrying pairs; it also has a Belpaire fire-box But here the likeness virtually ends. The cylinders of the Great Western engine are placed outside, those of the Lancashire and Yorkshire are inside the frames. The Great Western cylinders are 18 in. in diameter with a 30-in. piston stroke instead of being 19 in. in diameter with 26 in stroke, as in the case of the Lancashire and Yorkshire. Mr. Hoy's engine has a flat-topped dome; Mr. Churchward's No. 99 takes steam from a point in the greatest diameter of the tapering boiler. There is, therefore, a very marked difference even in the external appearance of the two locomotives. Moreover, the Great Western engine has considerably less heating surface than the other one, namely, 1,518 square feet. It will be

noticed that the exceptional length of piston stroke 30 in., introduced by Mr. Churchward in his 4-6-0 express class, is being used also in his large "Consolidation" goods engine and now in this new design of tank engine. Here, again, much interest will necessarily attach to the experiment whether the tractive advantage naturally secured by the increased length of stroke is or is not counterbalanced by any drawback. It is understood that the two ten-wheel six-coupled express engines, of which casual mention has already been made in passing, will come into regular main-line work ere long, having hitherto been undergoing a series of tests and trials. It has been found advisable, I understand, slightly to strengthen a few bridges which have not hitherto been subjected to the strain of a 69-ton engine running over them. There seems every reason to anticipate that these fine-looking engines with their six-coupled 6 ft. 8 in. driving wheels will give a very good account of themselves on the fast and heavy expresses between Paddington and Plymouth, should it be found judicious to run one engine through the whole distance of 246 miles.

## "La France"

There is no reason to doubt that the du Bousquet-de Glehn compound, No. 102, "La France," will acquit herself as well on this side of the Channel as on the other. The engine made her first trip on regular duty on the 2nd ult., taking the twelve noon semi-express from Paddington to Swindon, stopping at Reading by the way. The trip was in no sense a "trial" one. The engine was simply put, for her inaugural run, on a supremely easy task, namely to take a train, weighing well under 200 tons behind her tender, at an average start to stop speed of barely 44 miles an hour over an almost dead-level road. It is needless to say that this excessively light duty was performed without the smallest trace of effort. Reading was reached eight minutes in advance of booked time, and Swindon after five minutes delay, five minutes early. Up a 10-miles continuous slight gradient of 1 in 1,320 a speed of 67·2 miles an hour was attained and maintained. This, however, was manifestly child's play to a locomotive whose sister-engines have reached and sustained a speed of 75 miles an hour with a similar load up a gradient of 1 in 200. The sole point at issue is whether in virtue either of efficiency or of economy —the latter either in construction-cost, expense of maintenance, or consumption of fuel and lubricants—the French engine will prove to possess any material advantage over the very fine and efficient machines most recently designed and built for the Great Western Railway by Mr. Churchward, namely the "City" and "98" classes. That obviously can be determined satisfactorily only by practical experiment. In either case the reputation of neither class of engine will be prejudiced by the result. Both are admittedly admirable of their kind and thoroughly efficient. The real point at issue is whether it will be found desirable and advantageous for Britain to adopt the de Glehn system of compounding.

## Great Western Steam Pressure.

As a question has been raised as to the working steam pressure of the Great Western "City" class, I may state that the designer and builder, Mr. Churchward, informed me in writing that their working pressure is 180 lb. per square inch, although the boilers are constructed to carry 195 lb, should this at any time be required. The newest Great Western express engine No. 171 is to have the same pressure as "La France"—viz., 228 lb.

# THE
# LOTS ROAD     POWER     STATION.

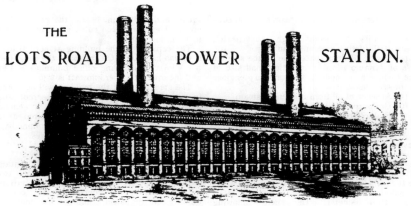

THE LOTS ROAD POWER STATION AS IT WILL APPEAR WHEN FINISHED.

BY

## HERBERT C. FYFE.

A short description of the power station which is being constructed at Lots Road, Chelsea  It
will be the largest in the world, and will have a most important bearing upon the development of
electric traction in the metropolis, while several of its engineering features are unique —ED.

THE great generating station which is now rapidly nearing completion in Lots Road, Chelsea, is noteworthy for so many reasons that a brief account of some of its most important features may be acceptable.

This immense power house will furnish current for the working of the Metropolitan District Railway and the three "Tubes" now under construction, and controlled by the "Underground Electric Railways Company of London, Limited," viz. :—

(1) Baker Street and Waterloo.

(2) Charing Cross, Euston. and Hampstead.

(3) Great Northern, Piccadilly and Brompton.

The total length of these lines is over sixty-three miles, the District Railway accounting for about forty. The work of laying the two conductor rails over the District system is now being rapidly pushed forward.

The Lots Road generating station is noteworthy for three reasons :—

(1) It will be the first great power house to employ steam turbines exclusively.

(2) It will be the largest electric traction station in the world.

(3) It will contain the largest steam turbines ever built.

The British Westinghouse Company are supplying these turbines. which will be ten in number, and will be of the Parsons type, with Westinghouse modifications. The speed will be 1,000 revolutions per minute. Mounted on the same shafts will be in each case a three-phase generator of 5,500 kilowatts. These will only have four field-magnet poles, and they will produce the energy at a potential of 11,000 volts, which is the highest pressure yet employed for traction purposes in Great Britain. The periodicity will be thirty-three-and-a-third per second.

The Westinghouse steam-turbine is of the Parsons parallel flow type, with such modifications as the experience of Westinghouse engineers in such work has suggested.

Owing to the absence of lubricating oil in the working cylinders, very high superheat can be used with advantage, and this, together with a high vacuum in the condensers, conduces to considerable steam economy.

The 5,500 kilowatt sets will occupy a floor space of 50 ft. long by 14 ft. wide.

The turbine itself is 29 ft. long over all, by 14 ft. wide and 12 ft. high : the rest of the space is occupied by the electric generators.

The principal advantages of the large rotary engine as compared with ordinary engines of the reciprocating type, may be summed up as follows :—

(1) Complete absence of vibration.

(2) Increased economy in steam and coal consumption.

(3) Facility for using high superheated steam with further economy resulting.

(4) Saving of floor space, due to the smaller size of dynamo possible, and the small size of the turbine as compared with ordinary engines of the same horse-power.

(5) Absence of oil from condensed steam.

(6) Uniformity of turning moment.

(7) Reduced cost of other classes of machinery.

(8) Reduced consumption of oil and stores.

With regard to the saving of floor space, it has been computed that if reciprocating engines were used in the Lots Road Station, running at seventy-five revolutions per minute, the diameter of the generators would have to be about 32 ft., whereas the fast-running dynamos actually to be employed will be little more than 9 ft. in diameter. The generators, owing to their high speed of rotation, are very much smaller than corresponding slow-speed machines would be.

### THE BUILDING AND ITS SITE.

The site comprises 3·67 acres of land, with a water frontage on the Thames and on Chelsea Creek of 1,100 ft., and a frontage on Lots Road, Chelsea, of 824 ft. There are two hundred and twenty concrete piers sunk to a depth of 35 ft. in the London clay.

The building is 453·5 ft. by 175 ft., and 140 ft. in height from the ground floor to the peak of the roof. The office building adjoining on the east, measures 81 ft. by 25 ft., and will have three floors, the lower of which forms the machine shops. The main building will have a self-supporting steel frame weighing about 5,800 tons. There will be four chimneys, each 19 ft. internal diameter and 275 ft. high; the foundations for these chimneys are 42 ft.

square and 34 ft. 6 in. below the ground floor level. There are 2,200 cubic yards of concrete in each foundation.

The capacity of the building at normal load is 57,000 kilowatts. On this basis the cubic feet per kilowatt (including office building) is 139, and the square feet per kilowatt is 1·36.

The steel frame of the building will be closed in with brick and terra-cotta; the roof and most of the floors will be concrete. In general details the building will be considered as a factory for the production of a commodity, and there will be no ornamental features.

#### BOILERS.

The south side of the building will contain eighty water-tube boilers arranged on two stories, a novelty so far as traction stations in this country are concerned, and carried directly on the steel frame of the building. Each boiler has 5,212 square feet of heating surface, and 672 square feet of superheating surface. The boilers will be piped in groups of eight, each group supplying the steam for one electric generating set and one feed pump, there being no steam connections between the several groups, except that a supplemental header at the east end of the building is connected to two groups. This header supplies the exciter engines, air compressors, house pump, etc. Chain grate stokers under each boiler have 83 square feet of surface.

Economisers with tubes 10 ft. long, and placed wider apart than the usual practice, are grouped behind the boilers, with the customary by-pass flues; 1,540 square feet of heating surface is provided for each boiler.

Boiler feeders are placed on the ground floor, and supply ring mains on both the boiler room floors.

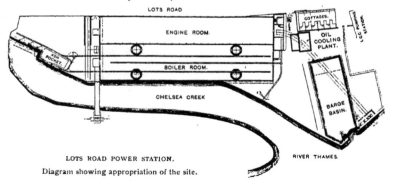

LOTS ROAD POWER STATION.
Diagram showing appropriation of the site.

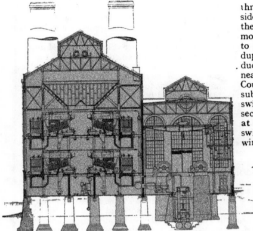

TRANSVERSE SECTION THROUGH ENGINE AND BOILER ROOMS.

### MAIN GENERATING SETS.

These consist of a horizontal turbine engine running at 1,000 revolutions per minute, and a three-phase generator wound for 11,000 volts 33⅓ cycles; there will be ten such sets (supplied by the British Westinghouse Company), wit floor space for one of half the size. The normal rating of each generator is 5,500 kilowatts, but they will carry an overload of 50 per cent. for two hours at practically the same steam consumption per kilowatt hour.

There will be four 125-kilowatts, 125-volt steam-driven exciter sets running at 375 revolutions per minute.

The switchboard (provided by the British Thomson-Houston Company) is carried on

three gallery floors extending across the north side of the engine-room, with returns across the east end. All high-tension switches will be motor operated, and the feeder system extending to the twenty-three sub-stations will be in duplicate. A line a mile long, of sixty-four ducts is completed to carry these feeders to the nearest point on the District Railway at Earl's Court, where they will diverge to the various sub-stations. The current used at the pilot switchboard will be low pressure from a secondary battery, and it will work the motors at the main board, which will operate the main switches. There will be about three miles of wires about the switchboard.

### THE CONDENSING SYSTEM.

This consists of vertical condensers each with 15,000 square feet of cooling surface; these are located in pits between the engine foundations. The circulating water is supplied by 66-in. pipes laid to the edges of the channel of the Thames. Each condenser has a 20-in. centrifugal pump; the duty of this pump is simply to overcome the friction of the pipes as the system is arranged on the syphonic principle; the top of the condensers being within 29 ft. of minimum low tide, and the circuit is closed. The intake and discharge mains are arranged for reversible flow.

The condensers are designed to work on the dry vacuum principle, the air pump and the water pump being separate. All the condenser pumps are electrically driven.

### FUEL SUPPLY.

Coal will be received on lighters in a tidal basin at the east end of the station, or by rail at an unloading point of the West London Extension Railway on the opposite side of

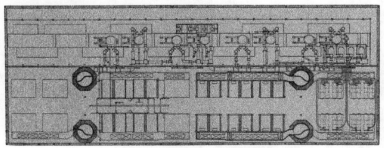

ARRANGEMENT OF STEAM AND EXHAUST PIPING.

Chelsea Creek. For unloading barge coal the basin is spanned by two travelling cranes, each working a one-ton grab ; the coal is weighed in the tower at one end of each of these cranes, and dropped on to a belt conveyor, thence by duplicate inclined elevators 140 ft. high .to the top of the building.

Rail coal will be taken from a hopper under the coal wagons by an inclined elevator to the top of the building at the opposite end. The distribution over the bunkers is by duplicate belt conveyors so arranged that the direction of travel of both belts can be reversed so as to handle coal coming in at either end.

The storage capacity of the bunkers is 15,000 tons. The daily consumption will reach 800 tons when the station is in full operation, and six of the largest river barges can be placed in the basin at each tide.

Ashes will be removed by an industrial railway worked by a storage battery locomotive ; two lines of rails will be laid under the ash hoppers on the ground floor. The ashes will drop into self-dumping buckets to be unloaded into barges by pneumatic hoists on the dock wall at the west end of the premises, or stored in an adjoining bin if no barge is available.

The capstans, barge basin gate mechanism, and many of the large valves in the building, will be worked by pneumatic motors.

The electric motors on the travelling cranes over the engines, as well as those on the oil switches, will be d.c. 125 volts. All other motors will be three-phase 220 volts ; most of the lighting will be on the latter circuit.

#### PROGRESS OF THE WORK.

Each of the four chimney stalks will be 275 ft. high. Each is 19 ft. internal diameter at the base, and the foundations are 42 ft. square, and 36 ft. deep. The chimneys are of brick, and are constructed on the custodis principle by the Alphons Custodis Chimney Construction Company. At the present time over 800 men are working at the foundations night and day. The foundations have to go down to the London clay, a depth of about 35 ft. The contractors for this part of the work are Messrs. Perry and Co. Like so many modern power stations, the building will consist of a steel framework

clothed with brick. The total quantity of steel to be used in the frame is about 5,800 tons. It was supplied by Messrs. Hein, Lehmann and Co., Düsseldorf. The British Westinghouse Electric and Manufacturing Company obtained the contract for the whole framework to be erected. They sublet the supply of the steel work to the German firm, and the erection to Messrs. Mayob and Haley. The machinery and boilers can be put in before the brickwork is finished, as the steel frame is absolutely self-supporting, and canvas sides can be supplied if necessary until the brickwork is finished.

The contract for the boilers and stokers has been given to Messrs. Babcock and Wilcox. An unusual feature in traction stations in this country is the fact that the boilers are to occupy two floors, one above the other. The contract for the big switchboard, which will control the supply of current to all the sub-stations on the District and its associated railways, has been let to the British Thomson-Houston Company, while Messrs. J. M. Simpson and Co. are supplying the condensers. The exciter engines are being built by Messrs. W. H. Allen and Co.

INTERIOR OF ONE OF THE CHIMNEY SHAFTS.

One of them is already running at the temporary power station previously referred to, and is furnishing light for the contractors, driving cranes, working an air compressor for pneumatic riveting, etc. There will be four of these engines, each of about 175 h.p., and each driving a British Thomson - Houston exciter dynamo. This plan: will be sufficient to furnish exciting current for the fields of the ten main generators. The compressed air apparatus has been supplied by the Consolidated Pneumatic Tool Company.

An interesting travelling electric crane was erected by Messrs. Jessop and Appleby, Leicester. It is of exceptional capacity. as it is capable of lifting thirty-five tons. The crane is erected close to the side of the creek, and it is used for lifting material and machinery from the barges and for handling these things on the site. It can lift the material from the barges and deposit it on platforms 19 ft. above in the power station.

In regard to the sub-stations, the actual positions of these on the District Railway are as follows : Whitechapel, Mansion House, Victoria, South Kensington, Earl's Court, Putney Bridge, Ravenscourt Park, Mill Hill Park, Hounslow, Sudbury. There will be approximately twenty sub-stations for the whole system, divided up between the "tubes' and the existing District Railway lines

The high tension feeder cables come from the British Insulated and Helsby Cables, Ltd., and the stoneware ducts from Doulton and Co., Ltd. The station was designed by Mr. James R. Chapman, General Manager and Chief Engineer of the Underground Electric Railway Company, and Mr. J. W. Towle is the engineer in-charge at the station.

Though the power station can easily be finished long before most of the new railways will be ready, every effort is being made to push on both the building and the machinery, because the sooner the station is finished the sooner will it be possible to work the District Railway itself electrically.

## OBITUARY.

THE sudden death from heart failure of Mr. W. G. McMillan has removed one of the best - known figures in the electrical world of the metropolis. It was occasionally our privilege to consult the late Secretary of the Institute of Electrical Engineers on matters concerning that body, and at such times one hardly knew whether most to admire the wonderful grasp of detail. which made him a force, professionally, or the personal characteristics of the man.

It was impossible to be long in Mr. McMillan's company without discovering that he was essentially a worker. With quiet deliberation and unfailing courtesy he seemed to carry in his own personality the very life of the institution, whose interests he had so much at heart. He was a man to whom those in search of information appealed with confidence, and rarely, if ever, were they sent empty away.

His indefatigable exertions on behalf of the members who joined the Italian trip are still fresh in mind, and little did we think, when Mr. McMillan conducted us over the new premises of the Institute shortly afterwards, that his useful work would so quickly come to an end.

Born in 1861, Mr. McMillan was educated at King's College School, and subsequently at King's College, where he ultimately joined the staff. Appointed by the Indian Government in 1888 as chemist and metallurgist to the Cossipore Ordnance Factories, near Calcutta, he also acted as Examiner in Chemistry to the University of Calcutta. During the time which elapsed between his return from India and his acceptance in 1897 of the post of Secretary to the Institution of Electrical Engineers. he turned his attention more particularly to the literature of electro-metallurgy with conspicuous success. He was an abstractor of the Society of Chemical Industry, a Fellow of the Chemical Society and of the Institute of Chemistry, a member of the Institution of Mining and Metallurgy, and in 1897 he was elected Vice-President of the South Staffordshire Institute of Iron and Steel Works' Managers.

*Photo by Elliott and Fry.]*
THE LATE MR. W. G. McMILLAN, F.I.C., F.C.S.,
Secretary of the Institute of Electrical Engineers.

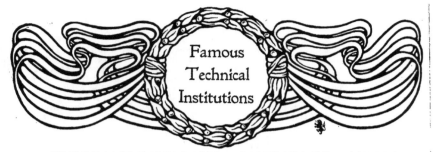

# Famous Technical Institutions

The third of a series of articles describing prominent technical institutions at home and abroad. The Massachusetts Institute of Technology was dealt with in the January number of PAGE'S MAGAZINE, and in the February issue a description of the Birmingham University was commenced. The author will conclude his remarks on this well-known centre of technical education in our April issue.—ED.

## II.—THE BIRMINGHAM UNIVERSITY.

BY

## C. ALFRED SMITH, B.Sc., A.M.I.E.E.

### PART II.

#### THE FOUNDRY AND FORGE.

WE mentioned in the previous article that there are two buildings adjacent to the power station. Of these, one is divided into the foundry and the smithery. The other is devoted to metallurgical work. The dimensions of the building for a section of the practical part of the engineer student's training is 100 ft. by 25 ft. and in height about 17 ft. It is a rectangular shaped building, with a dividing wall across the middle, leaving each shop 50 ft. by 25 ft. The equipment of the smithery will consist of twelve fires, and a motor-driven 3-cwt. power hammer fitted by Messrs. Thwaites Bros., of Manchester. In the foundry there will be a 2-ton cupola and brass-melting furnace, core stoves and an overhead crane. The building is quite finished, and the students will shortly be at work in it.

#### THE MAIN BUILDINGS.

Interesting as the equipment and construction of these three auxiliary buildings undoubtedly are, they pale into insignificance beside the huge laboratories and workshops which will form what is always called the " main build-ings." The general arrangement of these buildings may be seen from the illustrations. The magnitude of the scheme is probably not realised even in Birmingham itself, where the keenest interest is naturally taken in the new University. When it is remembered that the work and equipment of the buildings, already in hand, will cost almost half a million pounds, and that the maintenance, rates, and salaries of the staff, when once these buildings are completed—the probable time being next October—will probably be about or beyond £10,000 a year, some idea of the effect of this great outlay may be obtained. It is true that at present only four blocks are being built, the ultimate idea is to have ten blocks which it is estimated will cost upwards of a million pounds to build. The time for this to take place will, of course, depend solely upon financial considerations. However, the work in hand is large enough to overshadow anything else yet attempted in this country, although it is true that there are universities in America which are built on the same magnitude. The most imposing portion will be the central hall, which is 150 ft. by 75 ft., and will be used for great occasions, such as degree congregations and other public functions. Under this great hall there are the dining-rooms and kitchens for the students and staff.

When facing the great hall, the two blocks on the right are those for engineering students. In one of them there will be the electrical laboratory, 115 ft. by 50 ft., and at the end of

# THE TECHNICAL STAFF.

## ENGINEERING,
## MINING, AND     METALLURGY.

Mr. R. C. PORTER, M.Sc., A.M.Inst.C.E.    Mr. O. F HUDSON, A.R.C.S.    Mr. F. H. HUMMEL, A.C.G.I., A.M.Inst.C.E.
Prof. R. A. REDMAYNE, M.Sc., F.G.S.    Prof. BURSTALL, M.Sc., M.A., M Inst. C.E., M.I.M.E.    Prof. T. TURNER, M.Sc., A.R.S M.
Mr. C. A. SMITH, B.Sc., A.M.I.E E.    Mr E. H. ROBERTON, B.A.    Dr. D. K. MORRIS, Ph.D., A.M.I.E.E.

the block will be the hydraulic laboratory, 42 ft. by 110 ft. There will also be a large strength of materials laboratory for testing iron and steel. The general equipment of these are already decided upon, but as the contracts have not yet been let it is not possible to give a detailed description.

In the adjacent block there will be the hall of machines, the dimensions of which are 115 ft. by 50 ft. In the old buildings at Edmund Street there are some fine specimens of up-to-date machinery ; and the students there have turned out, among other work, during the past year, half a dozen lathes, and some motor-driven three-throw pumps. This machinery will be moved out to Bournbrook shortly. and will be largely added to. Mention should also be made of the spacious drawing office, which will accommodate at least 100 students at one time. There will also be a special room for taking blue prints, which work the students will have to do for themselves.

### A FOUR YEARS' COURSE.

It cannot be denied that the engineering course at Birmingham is a great experiment. It more nearly resembles the curriculum which the Admiralty have had in operation for the last fifteen years at Keyham College, Devonport, than any other training institution in England. It is based distinctly on opposite lines to that of the advocates of the "sandwich system," for it offers to teach the engineering student both theory and practice at the University. Its only weakness lies in the fact that, whereas the Admiralty trained engineer is brought fully into contact with the men in the shops, the Birmingham graduate will need yet another year, possibly as an improver at a small wage, in the shops in order to study that complex character, the British working man.

For the first two years all the engineering undergraduates take the same subjects. Great attention is given to mathematics, physics, chemistry and mechanical drawing. Elementary descriptive lectures in mechanical, civil and electrical engineering are also given. In the third and fourth years the student specialises in one of the above branches, and those who successfully pass all their examinations are,

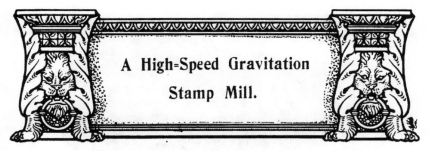

# A High=Speed Gravitation Stamp Mill.

BY

## EDGAR SMART.

A new type of stamp mill has recently undergone a thoroughly practical trial, under normal working conditions, at the Meyer and Charlton Mine in Johannesburg. The results obtained constitute such a marked advance on existing practice, that a description of the principal features of the improved stamp operating mechanism cannot fail to be of interest. The machine has been carefully and thoughtfully worked out in all its details, and, apart from its interest as a crushing mill, it is of considerable interest as a study in applied mechanics.—ED.

N a previous article on ore milling,* it was stated that the practical limits of speed for a gravitation battery of the ordinary type were about 95 drops per minute for a 9-in. drop, or 100 drops at 6 in. Reference was also made to some modified forms of stamp mechanism, which aimed at increasing the velocity of fall. An increase in the number of drops per minute, as well as in the energy of each blow, has been achieved by steam and pneumatic stamps, of which latter the Husband stamp is a well-known type, but neither of these forms has so far come into general use.

The battery about to be described differs from those mentioned above in that its object is not to accelerate the *fall* of the stamp, but to obtain a greater rapidity of action by means of a quicker *lift*. Therefore, although the cam is dispensed with, the machine is purely a gravitation battery, and, so far as the blow on the ore is concerned, it in no way differs from the ordinary type except that the stamps are of greater weight.

The invention is due, in the first place, to Mr. D. B. Morison, the well-known marine engineer and managing director of Richardsons, Westgarth and Co., Ltd., of Hartlepool; but in working out the details he has had the valu-

able assistance of Mr. D. A. Bremner, the managing director of the High-Speed Stamp Company, who personally superintended the trial of the new stamps at the Meyer and Charlton mine. Mr. Bremner has not only placed a large quantity of information concerning the machine at the author's disposal, but has also afforded every facility for the personal examination of the mill and its work, which is necessary in forming an independent opinion of its merits.

### CHIEF POINT AGAINST THE CAM.

The mill in its present form is the outcome of a practical trial of an earlier form and of more than five years' experimental work on a large scale by Messrs. Morison and Bremner, who have, during that time, carried out a number of systematic tests in crushing with cam stamps, as well as with the high-speed stamps. They have pointed out in communications to engineering societies, certain disadvantages inherent in the cam system, and which they have endeavoured to avoid in the new design. Of the various objections raised by them, the following appears to be the most serious and the most valid :—

The vertical motion of an involute cam, turning at a uniform velocity, is uniform throughout the lift, so that at the pick-up point the cam endeavours to lift the stamp immediately at full speed ; but this is impossible because the stamp, being at rest, cannot instantaneously

* Vol. I., page 315

Mortar Box Open.

Mortar Box Closed.

A BATTERY OF 1,600 LB. HIGH-SPEED STAMPS.

acquire the velocity of the cam. It is certain, therefore, that the motion of that part of the cam face, which is in contact with the tappet, must be retarded at the moment of impact. This involves either the yielding of the cam by bending or a local twisting of the cam shaft at each pick up, or, more probably, both of these effects together.

This objection is, of course, more theoretical than practical as regards cam stamps working under the conditions usual at the present time, for experience has abundantly proved that there is no difficulty in providing the necessary strength to bear these strains, large as they undoubtedly are. The importance of the point raised is in relation to the question of increasing the present limits of speed, because the objectionable strains above referred to, increase as the square of the cam velocity. As a mere matter of geometry, increased speed can be obtained by placing the stems farther from the cam shaft, so that the radius of the generating circle of the involute is larger, and the angle of rotation required for a given lift is consequently reduced. The same result can also be obtained in other ways, but in whatever way the vertical velocity is increased, the stresses on cams and shaft will be correspondingly intensified at the moment of pick up, and this may therefore be fairly considered as one of the chief factors which limit the possible speed of cam-driven stamps.

The discussion of this matter leads to a clear understanding of one desirable feature in a high-speed mill, namely, the gradual imparting of motion to the stamp. This, again, leads to the idea of an elastic cushion between stamp and lifting mechanism. which is only imperfectly provided in the ordinary battery by the yielding of the cam shaft and framing, which has already been mentioned.

### THE LIFTING MECHANISM.

In describing the new battery, it will be convenient to commence with the most novel and essential feature, namely, the device for picking up the stamp by means of a cylinder and piston, as shown in fig. 1. This figure differs slightly from the working drawing from which it was derived, in that all the parts necessary to the explanation of the action are shown in one sectional plan, whereas in the machine itself the various passages for air and water are distributed round the cylinder casting.

The cylinder $a$ reciprocates vertically, being actuated by a crank and connecting rod, placed above it. A top cover, $b$, carries the pin, c, of the connecting rod. At the bottom of the cylinder there is a packed gland, $d$, through which the piston-rod, $e$, passes. A port, $f$, in the side of the cylinder communicates with an annular water chamber, $g$, formed in the casting. Water is supplied to the chamber by the pipe, $h$, and overflows through the hole, $j$, to the escape pipe, $k$. The plain plug piston, $l$, which has no packing, is hollowed out to form an air vessel, $m$, and it has a central passage through it from $n$ to $o$. The size of the opening at $o$ can be varied when desired by the substitution of another vent plug, $p$, with a larger or smaller hole. This change can be made through the port door, $q$. A screwed plug valve, $r$, is provided for draining all the water from the cylinder when necessary. There are openings, $s\ s$, at the top of the cylinder, communicating with the water jacket to allow the free transfer of air inwards or outwards according to the relative motion of piston and cylinder.

### THE LIFT AND DROP.

Assuming that the stamp is resting upon the ore in the mortar box, and that the piston is therefore momentarily at rest, the cycle of operations is as follows :—

The cylinder completes its downward travel, and reaches the bottom of its stroke as shown in fig. 2. The port, $f$, then being open, the water can flow from the annular chamber, $g$, into the lower part of the cylinder, thus filling the whole space below the piston. It is evident that when the water level rises to the lower end, $n$, of the vent, the air in the vessel, $m$, is trapped. When the cylinder begins to rise, a small portion of the water is forced back through the port. which is, however, being rapidly closed, so that the pressure inside the cylinder quickly but gradually increases, and the trapped air is compressed, until the pressure is sufficient to overcome the weight of the stamp, which must then begin to rise. But, owing to its inertia, the stamp does not at once acquire the full velocity of the cylinder ; the air cushion on which it actually rests prevents by its elasticity any excessive strains in the mechanism, and the internal pressure continues to increase until piston and cylinder are moving upward together at the same rate. This state of things is shown in fig. 3. From the high-speed point of view, this gradual pick up is of the utmost importance as it allows of a very rapid lift without developing large initial stresses.

The cylinder velocity increases till about the middle of its upward stroke, and then gradually decreases. The acquired momentum of the stamp, and the expansion of the air beneath the piston. then carry the latter upwards faster

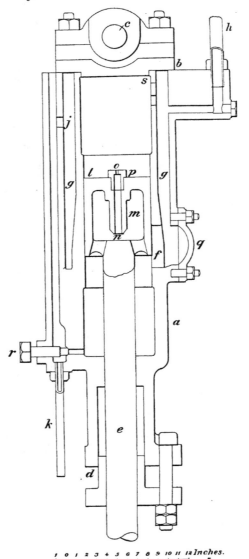

FIG. 1. DIAGRAM OF CYLINDER AND PISTON.

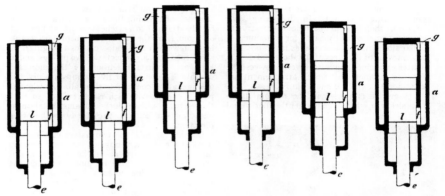

FIGS. 2 TO 7. SHOWING CYCLE OF OPERATIONS.

### CIRCULATION OF WATER.

The water is supplied to the cylinders from a small tank above the battery, whence it passes through a channel formed in one of the castings of the upper frame to a small fixed pipe which projects down into the pipe, *h*, shown in fig. 1, on the top of the cylinder. From the annular chamber the water overflows through the opening, *j*, to the pipe, *k*, which projects down into a fixed pipe of larger diameter, which carries the water to another channel formed in the frame, whence it passes to a small pump, which returns it to the tank. Thus, a slow but continuous circulation is maintained in such a manner as to continually aerate the water, to carry off the small amount of heat developed by the cushioning action, and to maintain a sufficient quantity at all times in the cylinder. A lubricant is mixed with the water to minimise friction and to obviate corrosion. In each of the telescopic pipes above mentioned, the inner and outer tubes do not actually touch each other, and therefore no wear or friction results from their relative motion.

### ADJUSTMENT FOR WEAR OF SHOES AND DIES.

For this purpose the arrangement shown in fig. 8 has been designed. The lower, tapered end of the piston rod, *e*, is cottered into a sleeve, *s*, about 30 in. long, which also fits on to the top end of the stem, *t*, and when in its lowest position abuts against a collar forged on the latter. The stem is held in the sleeve by a gib and keys, in the same way that the tappets are fixed to the stems in an ordinary battery. The distance from shoe to piston

can, therefore, be suitably adjusted by slipping the sleeve up or down on the stem and keying it in the required position.

### COMPENSATION FOR LOSS OF SHOE WEIGHT.

For the purpose of combining the above mentioned adjustment with compensation for loss of shoe weight, a number of washers are provided, each ½ in. thick, and shaped like a horseshoe, with the free ends turned down to lock into notches in the washer beneath. When the combined wear of a shoe and die amounts to about half an inch, the stamp is stopped, the keys slackened, the sleeve raised with a crowbar, and a washer slipped in between the flange of the sleeve and the collar of the stem. The sleeve is then dropped on to the washer, and the keys are tightened up. The insertion of the washers between the sleeve and stem flanges also prevents all chance of derangement through slipping of the stem within the sleeve. This simple means of compensating for the loss of shoe weight enables the weight, and therefore the crushing power, of the stamp to be maintained practically constant throughout the life of the shoe. The sleeves also allow of an alteration of the drop within certain limits when required, for by lengthening the distance between shoe face and piston, the latter is raised relatively to the cylinder, so that the pick-up of the stamp occurs later in the stroke, and the total rise and drop are lessened.

### STAMP ROTATION.

It is well known that in an ordinary battery the cam acting on one side only of the tappet imparts a more or less regular rotation to the stamp during each lift. In dispensing with the cam, therefore, it becomes necessary to provide some other means for obtaining a similar slow rotation to secure, as far as possible, uniform wear of the faces of the shoes and dies.

After experimenting with several devices, the following ingenious but simple arrangement has been adopted, which is shown in fig. 9 :—

A V-groove is turned in the bottom flange of the sleeve, $a$, which is consequently a horizontal

FIG. 8.

Showing piston rod, $c$, adjusting sleeve, $s$, stamp stem, $t$, and guide blocks, $u$.

V-pulley. About four feet away from this, at the back of the mill, there is a vertical V-pulley rigidly attached to a ratchet wheel, $b$, and carried on a pin, $c$, at the end of one arm of a bell crank lever, $d$, whose other end, $e$, carries a weight, $f$. A short rope belt, $g$, connects the two pulleys, and is kept at a constant tension by the aforesaid weight. The levers for five stamps are pivoted on one shaft, $h$, common to them all. A weighted pawl, $i$, engages with each ratchet wheel, so that the vertical pulley is locked against rotation in one direction, but can turn freely in the other direction.

Considering now the two halves of the rope, marked $n$ and $o$, it is obvious that as the stamp falls, the lower part, $n$, will be slackened, and the upper part, $o$, will be pulled towards the stem, thus tending to turn the vertical pulley in the direction of the arrow. The ratchet and pawl being arranged to allow of its motion in this direction, the vertical pulley will therefore be turned through a small angle. When the stamp ascends the upper part, $o$, of the rope is slackened, and the lower part pulled. As the vertical pulley is locked by the pawl against motion in this direction, the altered tensions in the two parts of the rope adjust themselves by rotating the horizontal pulley, and with it, of course, the sleeve and the stamp.

According to the direction of the teeth of the ratchet wheel, the rotation of the stamp can therefore be effected either during its rise or its fall. In practice, the former is preferred because it diminishes the frictional resistance during the lift.

This gear is placed behind the stamps under the platform, which gives access to the cylinders. The ropes of the turning gear can be clearly seen on the sleeves in fig. 8. In future it is proposed to use chains.

### THE CRANK SHAFT AND UPPER FRAME.

In this particular mill the five cranks have a throw of 4 in., and are so arranged that the stamps drop in the order, 1, 3, 5, 2, 4. The shaft can, of course, be designed for any order of drop, in the first instance, but this is afterwards unalterable. There is a bearing between each two cranks, as well as one at each end, thus making seven in all, and securing great rigidity, and adequate bearing surfaces. The bearings form part of a cast iron entablature, which is placed on the top of the battery frame, and bolted to two beams, which, in turn, are bolted to the king posts. The entablature and shaft are shown in fig. 10.

The centre line of the shaft is placed $3\frac{1}{2}$ in. in front of the centre line of the stamps, in order

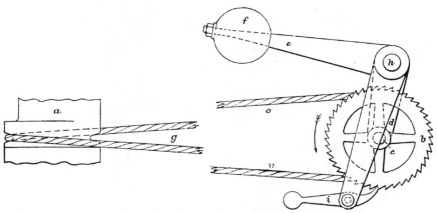

FIG. 9.   METHOD OF SECURING STAMP ROTATION.

that the connecting rod may be acting vertically during the loaded up stroke of the cylinder. This is, of course, only strictly true at one point of the stroke, but the arrangement ensures that the angular thrust against the cylinder guides shall be exerted chiefly during the down stroke when the cylinder is running free of any load.

As the five cylinders and connecting rods practically balance each other, no power is expended in raising any weight which does not contribute to the blow of the stamp.

A cast iron frame is bolted to the underside of the entablature at each end of it. These frames are connected by upper and lower cast iron beams, to which the vertical guide rods of the cylinders are fixed. Thus, the whole of the lifting mechanism is contained in one rigid metal structure, quite independent of the timber framing, so that no unexpected stresses due to the warping of the latter can occur. For the purpose of comparison with a cam battery, the mill may therefore be considered as consisting of three parts, viz.: (1) The self-contained lifting machine (including crank shaft, connecting rods, and cylinders), which corresponds to the cam shaft and cams of the ordinary mill. (2) The piston, piston rod, sleeve, and washers, which take the place of the tappet. And (3) the stem, head, shoe, die, mortar box, and timber framing common to both systems.

## STEMS, HEADS, ETC.

The stem only differs from that of a cam driven stamp in its larger diameter, decreased length, and in the forged collar previously mentioned, which abuts against the lower end of the adjusting sleeve. Its lower end is tapered, and fits into a conical hole in the head in the ordinary manner. The head and shoe are not only heavier than usual, but also of larger diameter, so that the area of the shoe and die are increased in proportion to the total weight of the stamp. The force of impact per unit of surface is, therefore, approximately equal to that of an ordinary 1,150-lb. stamp. The actual diameter in this case is 11 in., instead of the usual 9 in., and the stamps are spaced at 12-in. centres instead of 10 in. The heads, shoes, and dies are seen in position in fig. 11, and an ordinary shoe is shown standing on the plate at the right-hand side of the mortar box.

## MINOR DETAILS.

The propping-up gear deserves a word of notice in order to point out a difference caused by the absence of the cam. In an ordinary battery the stamp must be lifted higher than its normal lift, so that the finger can hold the tappet clear of the still-revolving cam. With the new battery the stamp merely requires to be held at or near the top of its stroke. This is accomplished by means of short forged steel fingers pivoted on pins carried in brackets mounted upon an I beam, which is bolted to the back of the king posts. These fingers engage with the flanges at the upper ends of the sleeves on the stamp stems, and they are fitted with safety pins to prevent accidental disengagement.

Provision is made for the continuous oiling of the crankshaft and connecting rod bearings by means of a central oil reservoir situated above the shaft.

### THE OPEN FRONTED MORTAR BOX.

The new mill at the Meyer and Charlton battery has one other unusual feature, which has, however, nothing to do with the question of speed, and is equally applicable to all types. This is an open-fronted mortar box, with a removable pressed steel front, which is shown in fig. 11. This front plate is dished out to stiffen it, so that it forms a rigid stay between the end walls of the box, to which it is fastened by six screws and nuts. The centre of it is attached by a link and swivel to a bar which can turn on a vertical hinge bolted to the framing. By this means it may be swung right out of the way, and so facilitate the work of cleaning up the mortar box or the renewing of stems, heads, etc.

FIG. 10. ENTABLATURE AND CRANK SHAFT.
Showing the seven bearings and five cranks.

### ACTUAL CRUSHING RESULTS.

During the final stage of the trials carried out at the Meyer and Charlton mine, the high-speed stamps were run continuously for four months, with only the stoppages customary in a stamp mill, and the following results were obtained :—

| | | |
|---|---|---:|
| Average weight of stamps | ... | 1,550 lb. |
| Average number of drops per minute | ... | 127 |
| Average height of drop | ... | 7¼ in. |
| Depth of discharge | ... | nil. |
| Screen mesh during July and August | ... | 500 |
| Screen mesh during September and October | | 400 |

Summary of 4 months' trial.

| | Hours run. | | Tons crushed. | Crushing rate per stamp per 24 hours. |
|---|---|---|---|---|
| | H. | M. | | |
| July | 576 | 25 | 1,013 | 8·43 |
| August | 645 | 25 | 1,203 | 8·94 |
| September | 568 | 25 | 1,054 | 8·90 |
| October | 599 | 55 | 1,167 | 9·33 |
| | 2,390 | 10 | 4,437 | 8.90 |

### CONVERSION OF EXISTING CAM STAMPS.

The suggestion has been made that existing mills could be converted by the substitution of, say, 1,400 - lb. high - speed stamps and the erection of the new mechanism on the old frame and foundations. The energy available calculated from weight and drop in several cases is as follows, the drop being 7½ in. :—

> A—1,000-lb. stamps, 95 drops, 59,375 ft.-lbs. per min.
> B—1,250-lb. stamps, 95 drops, 74,219 ft.-lbs. per min.
> C —1,400-lb. stamps, 126 drops, 110,250 ft.-lbs. per min.

A gain in crushing power — assuming this to be always proportional to the energy available — of 86 per cent., would therefore appear to be obtainable by conversion from A to C, and 49 per cent. by conversion from B to C, whilst, in either case, the cost would be but a fraction of that involved in the addition of the *pro rata* number of cam stamps.

### INSTALLATION OF NEW STAMPS.

It is not considered advisable to construct high-speed stamps of greater weight than 1,400 lb. on 10-in. centres, inasmuch as it would involve a sacrifice of first - class mechanical design, and therefore, that weight is the limit available for the conversion of existing cam stamp batteries, in which the width between king posts is generally five feet. These limits do not exist, however, in the case of new mills, and although 1,600-lb. stamps at 12-in. centres were decided upon as a satisfactory unit for introduction into mining practice, that weight is by no means the limit. In fact, it is asserted that considerable advantages will be derived from a still further increase of weight, and a corresponding reduction in the number of stamps required for a given output.

The writer has been informed that, to meet the conditions prevailing on the Rand, it is intended to standardise an 1,800-lb. high-speed stamp, having double the crushing power of the 1,300-lb. cam stamps now coming into vogue on that field. On this basis, 100 heads of high-speed stamps would be equivalent to 200 heads of cam stamps. It is claimed that, by reason of the large attendant savings in foundations,

FIG. 11. MORTAR BOX OF HIGH-SPEED STAMP MILL AT MEYER AND CHARLTON MINE, JOHANNESBURG.

Showing removable front plate, and heads, shoes, etc., in position.

building, framing, ore bins, battery tables, feeders, and line shafting, the initial cost of a mill on the Rand would be reduced by about 20 per cent. by the adoption of high-speed stamps.

# THE MANUFACTURE OF HEXAGON NUTS.

THE accompanying illustrations show a plant for the manufacture of hexagon nuts from 3 in. to 6 in. diameter from round stock. The steel bars are cut up into discs slightly in excess of the depth of the finished nut by a special cutting-off machine. The pieces as they leave this machine are in the form of circular blanks from each of which one nut is made.

They pass to the drilling machine, a substantial tool with three vertical spindles (fig. 1).

FIG. 1. THREE-SPINDLE DRILLING MACHINE.

The spindle heads are attached to a cross rail supported on two uprights. The spindles are balanced and have independent variable feed, with quick hand movement by lever and slow movement by hand-wheel. A stop is fitted disengaging the automatic feed immediately the hole is drilled through the nut blanks. Each spindle has a self-centring jig, into which the blanks are placed, and which is arranged to accommodate all sizes. The jig works on the table of the machine, and is provided with a set of steel bushes to suit the twist drills used for smaller nuts and the trepanning bits that are used for larger sizes. They are arranged so that the nuts may be slid between the gripping jaws and withdrawn without effort, a small platen being attached to each jig, having its surface just level with the bottom of the vice in which the drilling is done. A water-tank is fitted beneath the table and a small pump supplies a stream of lubricant to each drill.

The drilled blanks now pass on to the tapping machine, which is shown in fig. 2. This machine may be driven through either double or triple gearing, and in each case a range of five different speeds are available without alteration of belt and without stopping the machine This variation of speed is obtained by the nest of gears plainly seen in the headstock. The blanks are held by suitable means on the carriage, the motion of which is accurately governed by a stout guide-screw running in the centre of the bed, which is engaged by a double clasp nut. Change wheels to suit the pitch of thread are accommodated at the right hand end of the bed. The quick motion to the carriage is obtained by rack

# The Manufacture of Hexagon Nuts.

FIG. 3. NUT FACING AND CHAMFERING MACHINE.

and pinion movement actuated by a hand-wheel. The tap fits loosely in a square chuck on the spindle nose and feeds right through the nut. a stream of lubricant being directed on to the work during the operation.

The pieces are now brought to the nut-facing and chamfering machine (fig. 3), a special double geared lathe with carriage accommodating three tools and fitted with automatic cross transverse. A handle on the carriage controls the two rates of feed that are available and a drop-out worm-box disengages the feed at any desired point. First the roughing tool and then the finishing tool traverses over the face of the nut, after which the rate of feed is reduced by the handle change gear and the chamfering tool comes into operation. The nuts are held on their threads by means of a collapsible mandrel which projects from the spindle nose. This is controlled by a long hand-lever working off a fulcrum at the tail-end of the spindle.

The nut is slipped into position and the mandrel is then expanded to grip the threads inside the nut. It screws itself home again against a rocking washer, and operations commence immediately. As soon as the work is done, the mandrel collapses by means of the controlling hand-lever and the nut is slipped off. The mandrel is in the form of four loose jaws, which are threaded to suit the nut, and a set of jaws is supplied for each size of nut. When the nut is faced on both sides and chamfered, it passes to the hexagon milling machine, shown in fig. 4. Two or more nuts are placed on the mandrel, which has a regular circular feed by worm and wheel. Behind this is situated a cutter headstock, which is mounted on a slide so as to be capable of a reciprocating motion to and from the work. This reciprocating motion is governed by a pair of cams at the rear of the machine, and in the case of hexagon nuts, are connected to the work mandrel by change gears, so that the cams make six revolutions whilst the nuts turn once.

FIG. 2. NUT TAPPING MACHINE.

The cams are so shaped as to give the requisite motion to the cutter head enabling the milling cutter to remove the material and form the side of a nut correctly. Side after side is thus formed without an interval and without attention from the operator until the nuts have made one complete revolution, when the self-acting feed becomes disengaged automatically and the nuts are completed. The machine may be set instantly for doing any size of nut without change of cams, and the work is turned out most expeditiously.

The whole plant is efficient in its operations, and it has been proved that nuts can be produced complete and bright all over at the cost of making rough forgings from which these nuts are usually manufactured. For the photographs illustrating the above

FIG. 4.   HEXAGON MILLING MACHINE.

plant we are indebted to the makers—Messrs. John Holroyd and Co., Ltd., of Milnrow, near Rochdale.

---

The first of the two refuse destructors erected by Messrs. Meldrum Bros , Ltd., for the city of Johannesburg is now being started to work and the second will be speedily completed   The company are putting down plants at the present time for the following towns .  Preston, Holyhead, Watford, Epsom, Ayr, Hyde, Kettering, Glasgow, Todmorden, Ossett, Manchester Corporation, Barnes, Walthamstow, etc. These will furnish power for electric lighting, traction, sewage pumping and other municipal purposes.

The British Westinghouse Electric and Manufacturing Company, Ltd , are supplying two gas engines and auxiliaries to the Tiverton Corporation, through Messrs. Frank Suter and Co., of Berners Street, W. These engines are of 125 b.h p. each, and are of the three-cylinder (15 in. by 14 in.) type.  The Chesterfield Corporation have recently ordered from the Westinghouse Company a new 400-kilowatt steam dynamo, together with switchboard and auxiliary apparatus, a Babcock and Wilcox boiler and a Worthington jet condenser.

Messrs. Thomas Piggott and Co., Ltd., Atlas Works, Birmingham, inform us that they have removed their London offices from 14, Great St. Thomas Apostle to 63, Queen Victoria Street.

An order has lately been placed by the General Electric Company, of the United States, with Messrs. Mather and Platt, Ltd., for fourteen of their special type centrifugal pumps for high lifts.  These pumps will be driven by General Electric Company motors connected through a flexible coupling, and are intended for use in two gold mines in India ;  six pumps will be connected in series in each mine ; the seventh being a spare.

The Stirling Company, of U.S.A., advise us that they have now opened a London office at 53, Victoria Street, Westminster, S.W., to enable them to deal more efficiently with their London and South of England business. This office will be under the charge of Mr. G. Graham. The head office of the company will be at 53, Deansgate Arcade, Manchester as heretofore.

JAPANESE ARTILLERY IN ACTION.
Covering an advance.

# THE WAR BETWEEN RUSSIA AND JAPAN.

BY

## N. I. D.

In view of the concentration of interest upon events in the Far East, our naval correspondent is confining his attention this month to the war. His remarks cover the period between the commencement of hostilities and February 24th. For many of the illustrations we are indebted to *The King and His Navy and Army.*—ED.

WITH dramatic suddenness the negotiations between Russia and Japan with reference to the irrespective positions in the Far East, were broken off at the beginning of last month ; but with even more dramatic suddenness did the Japanese readjust the balance of naval power which for some months has been slowly but surely tending to Russia's favour.

With the diplomatic side of the question I am not here concerned. For an understanding of the situation it is merely necessary to bear in mind the disposition of the forces of the combatants in the early days of February. It was on the 5th of that month that the Mikado's Government intimated to St. Petersburg that, their patience being exhausted, they must terminate negotiations with words, and appeal to the sword. It has since been learned that

on or about that day the Japanese fleet, with the exception of a few small cruisers, was ordered to assemble at Sasebo, under the command of Admiral Togo. It is believed that six battleships, the *Mikasa*. flagship, *Asahi*. *Fuji*, *Yashima*, *Shikishima*, and *Hatsuse*, with six armoured cruisers, the *Asama*, *Yakumo*, *Azuma*, *Idzumo*, *Iwate*. and *Tokiwa*, and at least eight protected cruisers with nineteen torpedo boat destroyers concentrated there.

The Russians, on the contrary, in defiance of all recognised tenets of strategy, distributed their fleet between three points, Port Arthur, Vladivostok. and Chemulpo, knowing well that in the event of war, the vessels at the northern port could hardly hope to effect a junction with those in the Yellow Sea. At Port Arthur there were gathered the seven battleships, *Retvisan*,

A GENERAL VIEW OF PORT ARTHUR.

Cesarevitch, Palla P
Peresviet, and Pobiem
cruiser Bayan and t
Novik Askold. Durin
Vladivostok there th
the Gromovoi, Rossia, &
could easily have left t
time, as the harbour
And at Chemulpo que
and the old gunboat A
much use such for
in the combined flee
that for some cause
has ceased to be xa
whole of the Russian
exception of the vess
to leave the inner wate
29th, to avoid being as
forming, and to be

Such, then, was the
fleets on the first da
state of war ma b
existence. It would b
Admiral Alexieef was i
position his fleet was
it, at least Vice Adm
steps were taken to
safe place, the respo
responsible acted se ass
burg or not is never si
result was the same
in an exposed post
were taken, such as w
by common prude
an easy prey.

For Admiral Tg
Yellow Sea, followin
armoured cruisers
protected cruisers
and Akashi, six
transports. It may se
were strategically bas
transports before ha
of the sea. Had the
been a torpedo flotilla
of events they should
Port Arthur fleet na
its moorings,

(2,2)

# The War between Russia and Japan.

*Cesarevitch, Poltava, Petropavlovsk, Sevastopol, Peresviet,* and *Pobieda,* with the armoured cruiser *Bayan,* and the small cruisers *Boyarin, Novik Askold, Diana,* and *Pallada.* At Vladivostok there were four powerful cruisers, the *Gromovoi, Rossia,. Rurik,* and *Bogatyr,* which could easily have left for Port Arthur at any time, as the harbour was kept clear of ice. And at Chemulpo were the cruiser *Variag,* and the old gunboat *Korietz,* neither of them of much use singly, but important integral factors in the combined fleet. It is further to be noted that for some cause or another Port Arthur has ceased to be ice free in the winter, and the whole of the Russian fleet there, with the exception of the *Sevastopol* in dock, was forced to leave the inner waters hurriedly on January 29th, to avoid being locked up in the ice then forming, and to lie outside the breakwater.

Such, then, was the position of the opposed fleets on the first day (February 6th), that a state of war may be said to have been in existence. It would be foolish to suppose that Admiral Alexeieff was unaware of the dangerous position his fleet was in. If he did not know it, at least Vice-Admiral Stark did. But no steps were taken to put the fleet in the only safe place, the open sea. Whether those responsible acted on instructions from St. Petersburg or not is never likely to be known. The result was the same. A powerful fleet was left in an exposed position, no apparent steps were taken, such as would seem to be indicated by common prudence, and the Japanese found an easy prey.

For Admiral Togo had steamed into the Yellow Sea, followed by Admiral Uriu with the armoured cruisers *Asama* and *Chiyoda.* the protected cruisers *Takachiho, Naniwa, Suma,* and *Akashi,* six torpedo boats, and several transports. It may be urged that the Japanese were strategically foolish to send out their transports before having established command of the sea. Had the two vessels at Chemulpo been a torpedo flotilla, as in the correct course of events they should have been, and had the Port Arthur fleet not remained cooped up at its moorings, the earliest engagements of the war would in all probability have made different reading. But the very daring and success of the manœuvre shows that the Japanese had not miscalculated the probable action of their opponents.

In the forenoon of February 8th, according to the Japanese official information, Admiral Uriu's squadron met the *Korietz* as she was coming out of port, at Chemulpo. The *Korietz* took up an offensive attitude, and fired upon the Japanese torpedo boats, which replied with torpedoes, none of which took effect. It is not stated that Admiral Togo's fleet was then in sight. In all probability it was not. At any rate. neither of the Russian vessels could get out to carry the news to Port Arthur, and the consequence was the *débâcle* of midnight, February 8th to 9th. It is best described in the official despatch sent by Admiral Alexeieff to the Emperor. In its very baldness it is more significant than columns of description :—

I most respectfully inform your Majesty that at about midnight on the night between February 8th and 9th Japanese torpedo boats made a sudden attack by means of mines [torpedoes] upon the squadron in the outer roads of the fortress of Port Arthur, in which the battleships *Retvisan* and *Cesarevitch* and the cruiser *Pallada* were damaged. An inspection is being made to ascertain the character of the damage. Details are following for your Majesty.

When the details came to be known it was at once obvious that the attack had been far more successful than even the first intimation had led us to believe. The *Cesarevitch* was damaged in her steering compartment, the *Retvisan* below the water line, and her pumping gear put out of order; while the damage to the *Pallada* was amidships, " near the engines." And since, up to the time of writing, it has not been found possible to get the *Pallada* into dock, it looks very much as if that " near the engines " was to read " in the engine-room."

The following day the Japanese began a bombardment of Port Arthur, and in the course of the fight, which lasted about an hour, the battleship *Poltava* and the cruisers *Diana, Askold,* and *Novik* were each damaged on the water line. The Russian losses were thirteen killed and fifty-four wounded. Admiral Togo, in a dispatch, dated at sea, February 10th. states that his losses were four killed and fifty-four wounded, and the damage to the ships was slight. Subsequently, he reported the ships

*(Continued on page 251.)*

15 A

THE RUSSIAN BATTLESHIP "PERESVIET."

Displacement, 12,674 tons; I.H.P., 14,500; speed, 18 5 knots; heaviest armour, 9·5-in.; heaviest gun, 10-in.

**THE NEW RUSSIAN FIRST-CLASS BATTLESHIP "CESAREVITCH," TORPEDOED AT PORT ARTHUR BY THE JAPANESE.**

Displacement, 13,100 tons ; I.H.P., 16,300 ; speed, 19 knots.

ARMOUR PROTECTION : Complete water-line belt of 10-in. armour, tapering to 4 in. at bow and stern, and to 7 in below the water. A 6-in. upper belt above reaching to main deck, tapering to 2·5 in. at ends. Turrets for heavy guns, 10-in. : secondary turrets for 6-in. guns, 7 in. ; ammunition hoists, 5 in. ; armoured deck, 4 in. on slopes, 2 in. on flat ; conning tower, 10 in., with 5 in. tube.

ARMAMENT : Four 12-in. 40-calibre guns in pairs in turrets, one forward and one aft ; twelve 6-in. 45-calibre Q.F. guns in pairs in small turrets ; twenty 12-pounder, twenty 3-pounder, and six 1-pounder Q.F. guns, with six torpedo tubes, two submerged forward.

THE RUSSIAN BATTLESHIP "RETVISAN."

Built in Philadelphia. Length, 376 ft. ; displacement, 12,700 tons ; speed, 18 knots ; heaviest gun, 12 in.
The "Retvisan," "Cesarevitch," and "Pallada," were torpedoed in the first engagement.

THE RUSSIAN BATTLESHIP "POBIEDA."

Built at St. Petersburg. Length, 435 ft.; displacement, 12,674 tons ; speed, 18 knots ; heaviest gun, 10 in.

THE JAPANESE ARMOURED CRUISER "ASAMA."

Displacement, 9,750 tons; heaviest gun, 8-in.; thickest armour, 7-in., I.H.P., 13,000 : nominal speed, 21·5 knots. Built by Armstrong, Whitworth and Co., Ltd.

THE RUSSIAN ARMOURED CRUISER "ASKOLD."

Displacement, 6,500 tons; heaviest gun, 6 in.; nominal speed, 23 knots. Damaged on water line.

THE JAPANESE ARMOURED CRUISER "YAKUMO."

Built at Stettin in 1899. Length, 426 ft. ; displacement, 9,850 tons ; speed, 20 knots ; heaviest gun, 8 in.

THE JAPANESE BATTLESHIP "MIKASA."

Displacement, 15,200 tons ; heaviest gun, 12-in. ; thickest armour, 14-in. ; I.H.P., 15,000 ; nominal speed, 18 knots.
[The "Mikasa" and the "Asahi," both built in British yards, may be termed the premier battleships
of the Mikado's navy.]

A DETACHMENT OF RUSSIAN IRREGULARS.

Ready for service in the East.

(*Continued from p. 243.*)
da maged were the *Mikasa, Fuji, Yakumo,* and *Iwate.*

At about the same time as the bombardment of Port Arthur was proceeding, Admiral Uriu was engaging the *Variag* and *Korietz* outside Chemulpo. The result of the action was that both Russian ships returned to harbour badly damaged, where they were blown up by their crews, who escaped to the foreign vessels in the port. The Russian losses were 47 killed and 62 wounded. No casualties were reported from the Japanese side.

The torpedo flotillas which accompanied Admiral Togo were divided into two groups, one of which carried out the successful attack at Port Arthur, and the other of which appears to have gone to Dalny and drawn blank. This latter flotilla was not, however, quite deprived of an opportunity to distinguish itself, as on the night of February 14th—15th it was despatched to Port Arthur to complete the work begun six days earlier. Owing, however, to a blinding snowstorm, only two vessels succeeded in reaching the Russian Fleet, the torpedo boat destroyers *Asagiri* and the *Hayatori.* One of these succeeded in torpedoing an enemy, believed to be the *Boyarin,* but also reported to be the *Bayan.* Either way, the exploit was a remarkable one, and in no way less useful than that which heralded the war.

Of the Vladivostok squadron, it is sufficient to note that it has been cruising in the Northern Sea of Japan, sunk an unarmed merchantman on February 11th, and threatened to bombard Hakodate, an unprotected port, both of them criminal actions, which will do little to advance the progress of the war, and much to damage Russia's cause with the other powers. On February 11th also the mining transport *Yenessei* struck one of her own mines, and was blown up with great loss of life.

On February 16th, the new Japanese cruisers *Kasuga* and *Nisshin* arrived at Yokosuka, the English officers who had assisted to navigate them before war was declared, being received and decorated by the Mikado personally.

On February 24th, the important news was circulated that Admiral Virenius and the squadron which had been lying at Djibutil for some weeks was ordered home.

The effect of the action at Port Arthur on the naval situation will be better understood from a study of the following tables :—

## RUSSIA—FLEET IN THE FAR EAST.

| Type. | Vessel | Date of Launch. | Displacement. Tons. | Speed. Knots. | Coal Capacity. Tons. | Principal Armament. | Weight of Broadside. lb. |
|---|---|---|---|---|---|---|---|
| b | *Peresviet* | 1898 | | | | | |
| b | *Pobieda* | 1900 | 12,674 | 18·5 | 2,056 | 4 10-in. 11 6-in., 20 3-in | 2,672 |
| b | *Ossliabia** | 1898 | | | | | |
| b | *Poltava* | | | | | | |
| b | *Sevastopol* | 1894 | 10,960 | 17·5 | 1,050 | 4 12-in., 12 6-in. | 3,367 |
| b | *Petropavlovsk* | | | | | | |
| b | *Retvisan* | 1900 | 12,700 | 18 | 2,000 | 4 12-in., 12 6-in , 20 3-in.. | 3,434 |
| b | *Cesarevitch* | 1901 | 13,100 | 19 | 1,350 | 4 12-in., 12 6-in., 20 3-in. | 3,516 |
| a.c. | *Bayan* | 1899 | 7,800 | 22 | 1,100 | 2 8-in.. 18 6-in., 20 3-in. | 952 |
| a.c. | *Gromoboi* | 1899 | 12,336 | 20 | 2,500 | 4 8-in., 16 6-in., 20 3-in. | 1,197 |
| a.c. | *Rossia* | 1896 | 12,200 | 20 | 2,500 | 4 8-in., 16 6-in. | 1,348 |
| a.c. | *Rurik* | 1892 | 10,940 | 18 | 2,000 | 4 8-in., 16 6-in., 6 4·7-in. | 1,345 |
| a.c. | *Dmitri Douskio* | 1883 | 5,893 | 15 | 400 | 5 6-in., 6 4·7-in. | 444 |
| p.c. | *Askold* | 1900 | 6,500 | 23 | 1,100 | 12 6-in., 12 3-in. | 772 |
| p.c. | *Variag* | 1899 | 6,500 | 23 | 1,250 | 12 6-in., 12 3-in. | 510 |
| p.c. | *Pallada* | | | | | | |
| p.c. | *Diana* | 1899–00 | 6,630 | 20 | 1,430 | 8 6-in., 22 3-in. | 632 |
| p.c. | *Aurora** | | | | | | |
| p.c. | *Boyarin* | 1901 | 3,200 | 22 | 650 | 6 4·7-in. | 180 |
| p.c. | *Novik* | 1900 | 3,000 | 25 | 600 | 6 4·7-in. | 180 |
| p.c. | *Bogatyr* | 1901 | 6,750 | 23 | 1,430 | 12 6-in., 12 3-in. | 872 |

* On passage out.

### JAPAN—FLEET IN THE FAR EAST.

| Type. | Vessel | Date of Launch. | Displacement. Tons | Speed. Knots. | Coal Capacity. Tons. | Principal Armament | Weight of Broadside. lb. |
|---|---|---|---|---|---|---|---|
| b | Hatsuse | | | | | | |
| b | Asahi .. | 1899 | 15,000 | 18 | 1,400 | 4 12-in., 14 6-in , 20 3-in... | 4,240 |
| b | Shikishima | | | | | | |
| b | Mikasa .. | 1900 | 15,200 | 18 | 1,500 | 4 12-in., 14 6-in., 20 3-in... | 4,250 |
| b | Yashima | 1896 | 12,300 | 18 | 1,100 | 4 12-in , 10 6-in., 20 3-in... | 4,000 |
| b | Fuji .. | | | | | | |
| a.c. | Tokiwa | 1898 | 9,750 | 21·5 | 1,200 | 4 8-in , 14 6-in , 12 3-in. .. | 3,568 |
| a.c. | Asama .. | | | | | | |
| a.c. | Yakumo | 1899 | 9,850 | 20 | 1,100 | 4 8-in , 12 6-in., 12 3-in. .. | 3,368 |
| a.c. | Azuma .. | 1899 | 9,436 | 21 | — | 4 8-in , 12 6-in., 12 3-in. .. | 3,368 |
| a.c. | Idzumo | 1899–00 | 9,800 | 24·7 | 1,600 | 4 8-in., 14 6-in., 12 3-in. .. | 3,568 |
| a.c. | Iwate | | | | | | |
| p.c. | Takasago | 1897 | 4,300 | 24 | 1,000 | 2 8-in., 10 4 7-in., 12 3-in. | 800 |
| p.c. | Kasagi .. | 1897–98 | 4,784 | 22·5 | 1,000 | 2 8-in., 10 4 7-in., 12 3-in. | 800 |
| p.c. | Chitose .. | | | | | | |
| p.c. | Itsukushima | | | | | | |
| p.c. | Hashidate | 1890–91 | 4,277 | 16·7 | 400 | 11 2·5-in , 11 4·7-in. .. | 1,260 |
| p.c. | Matsushima | | | | | | |
| p.c. | Naniwa | 1885 | 3,727 | 17·8 | 800 | 2 10·2-in., 6 5·9-in. .. | 1,196 |
| p.c. | Takichiho | | | | | | |
| p.c. | Yashima | 1892 | 4,180 | 23 | 1,000 | 4 6-in, 8 4·7-in. .. .. | 780 |
| p.c. | Akitsushima | 1892 | 3,150 | 19 | — | 4 6-in, 6 4 7-in. .. .. | 780 |
| p.c. | Niitaka | 1902 | 3,420 | 20 | 600 | 6 6-in. .. .. .. | 920 |
| p.c. | Tsushima | | | | | | |
| p.c. | Suma .. | 1895 | 2,700 | 20 | 200 | 2 6-in., 6 4 7-in. .. .. | 385 |
| p.c. | Akashi .. | | | | | | |

The immediate effect of the reverse to the Russians in the Far East was to set rumour active with regard to the Baltic and Black Sea fleets. The Black Sea fleet, it is quite permissible to leave altogether out of consideration in reference to the situation in the Far East. For not only are the vessels too old to be of much use out there, they are also of too small coal capacity to render it practicable to send them on the journey, even could permission be obtained to let them pass the Dardanelles. Much the same objection applies to the Baltic fleet, as will be seen from the following table :—

### RUSSIA—FLEET IN THE BALTIC.

| Type. | Vessel. | Date of Launch. | Displacement Tons. | Speed. Knots. | Coal Capacity. Tons. | Principal Armament | Weight of Broadside. lb. |
|---|---|---|---|---|---|---|---|
| b | Sissoi Veliky .. | 1894 | 8,800 | 16 | 800 | 4 12-in., 6 6-in. .. .. | 3,099 |
| b | Navarin .. | 1891 | 9,476 | 16 | 1,000 | 4 12-in., 8 6-in .. .. | 3,172 |
| b | Nicolai I. | 1889 | 9,700 | 15·9 | 1,200 | 2 12-in , 4 9-in , 8 6-in. .. | 2,292 |
| b | Alexander II. .. | 1889 | 9,900 | 16·5 | 1,200 | 2 12-in , 4 9-in , 8 6-in. .. | 2,292 |
| b | Borodino | | | | | | |
| b | Alexander III. | | | | | | |
| b | Orel .. .. | 1901–02* | 13,516 | 18 | 1,250 | 4 12-in , 12 6-in., 20 3-in. .. | 3,456 |
| b | Kniaz Suwaroff | | | | | | |
| b | Slava .. .. | | | | | | |

* These vessels are completing, and will not be ready to leave till the late spring.

In all these tabular statements I have only taken into account the larger vessels. Small craft would be practically useless for the Russian purpose out there, even if they could get there, which is more than doubtful.

PART OF THE GENERATING PLANT. BATH TRAMWAYS.

# THE BATH TRAMWAYS.

## A Description of the System, with a Brief Account of the Inauguration.

### BY A STAFF CORRESPONDENT.

A SMALL enclosed area — a mere nucleus within an arm of the River Avon — represented the Aqua Sulis of the Romans. Modern Bath is an altogether different affair. Since the time of Beau Nash, whose visitors arrived in stage coaches, the city has steadily encroached upon its beautiful surroundings, until picturesque Bathford, on the extreme east, seems anxious to join hands with the residential suburb of Weston, and the more industrial district of Twerton on the other side of the map. Thanks to the new tramways, this has actually come to pass.

The site of the City is said to be of volcanic origin, and, as is well known, lies in a hollow, the springs rising in the heart of the city within a few yards of

CAR SHED AND OFFICE BUILDINGS.

MAIN SWITCHBOARD.  BATH ELECTRIC TRAMWAYS.

as engineers.  The main contract for the tramways was awarded to Mr. Charles Chadwell, of 20, Victoria Street, London, S.W., and the track, overhead equipment, and the feeder cables, were laid and fitted by him.

To the British Westinghouse Electric and Manufacturing Company was entrusted the erection and equipment of the generating station (which has a capacity of 675 kilowatts), a car shed, for forty cars, office buildings, and the fitting of forty Milnes cars with Westinghouse 49 B. 30 h.p. motors, No. 90 controllers, and the Westinghouse magnetic brake.

the abbey.  The surrounding country presents difficult gradients, and many an invalid must have cast longing glances at the neighbouring heights.  In future the trams will allow visitors to make a choice of two climates—that of the city of Bath and the more bracing air of the suburbs, while students will be able to vary the pursuit of Roman antiquities with delightful excursions into the country.

The trams have been running since the beginning of the year, and have already proved themselves well - nigh indispensable. The system extends to within three or four miles of the Bristol system, so that the ultimate interconnection of the two is by no means improbable.

The Bath Electric Tramways, Ltd., originated the project, and Messrs. George Hopkins and Son, of Clun House, Surrey Street, London, W.C., Messrs. Harper Bros. and Co., of 13, St. Helen's Place, London, E.C., and Mr. R. D. McCarter, of London, acted

ONE OF THE 200 K.W. WESTINGHOUSE DIRECT-CURRENT TRACTION GENERATORS.

NEW TRAMS PASSING THE GUILDHALL. BATH ABBEY IN THE BACKGROUND.

The car shown in the foreground has to encounter some difficult gradients, its destination being the heights of Combe Down.

IN THE BOILER HOUSE.

public on Saturday, January 2nd.

At the invitation of the chairman of the company (Sir Vincent Caillard) and the directors, an inaugural luncheon was held at the Guildhall, Bath, on Saturday, February 13th. An inspection of the power house was first made. Lady Sivewright started one of, the cars, and formally declared the tramways open, being subsequently presented with a silver box as a memento of the occasion. Trips on reserve cars had been arranged, but the inclemency of the weather made everybody anxious to reach the Guildhall as soon as possible.

The following particulars of the system are from an official description, which we owe to the courtesy of the British Westinghouse Company :—

Mr. R. D. McCarter is the general manager and engineer for the tramway company, and during the building and equipment of the works, and since they started running, has made his headquarters at Bath.

It is greatly to the credit of the Westinghouse Company that they carried out their work in less than 7½ months, this being — we believe — a record time for the erection and equipment of a traction station of this capacity. The ground was broken for the building foundations on May 1st last, and the Board of Trade inspection took place on December 12th. The system was opened to the

## GENERAL OUTLINE OF THE SCHEME.

The length of the lines now in operation comprises nearly 16 miles of single track, the total length of the

A BOOSTER SET. BATH ELECTRIC TRAMWAYS.

routes being 12¼ miles. The gauge is the standard 4 ft. 8½ in. The steepest gradient is 1 in 11, and is 4 chains long, there being also five short lengths of 1 in 12 ; the worst gradient, however, is 1 in 14, 18 chains long. The sharpest curve is 37 ft. radius, and there are four less than 40 ft., and ten less than 50 ft. Four routes radiate from the centre of the city, two of them with short spurs, and one with an elongated loop. The construction of another mile of route is authorised, and will go to form two loops in the city itself.

The buildings are situated on the banks of the Avon, and consist of a boiler-house, engine-house, and car-shed. They are similar in character to those of the power station erected by the British Westinghouse Company for the working of the Mersey Railway. The boiler-house is 93 ft. long and 49 ft. wide ; the engine and generator house being the same length, but 2 ft. wider. The car-shed is about 130 ft. long and 82 ft. wide. Over the street end of the car shed are various rooms for offices etc., while at the rear end is a large basement which is used as a repair shop. Eight tracks run the full length of the shed, an inspection and repair pit 30 ft. long being built in each track. The walls of the building are of brick. The engine room walls inside are of glazed brick to a height of 10 ft. and the roof trusses and the purlins for slating are entirely of steel. The iron and steel constructional work was done by Messrs. Lysaght and Co., and the glazing by Messrs. Mellowes, as sub-contractors to the British Westinghouse Company.

The engine foundations extend some 25 ft. to 30 ft. below the floor of the engine room, the floor being of concrete finished with cement. The chimney shaft is 140 ft. high, 13 ft. 6 in. diameter at base, and 7 ft. in diameter at the top under the cap. It is of sheet steel, lined inside with brick to a height of 60 ft., and was built by the British Westinghouse Company. The chimney rests on a bed of concrete and brickwork. Though self-supporting, it is held down by bolts as an extra precaution, and is fitted with a permanent ladder.

#### THE STEAM PLANT.

This comprises three Babcock and Wilcox boilers, each with a heating surface of 3,140 square feet, making a total of 9,420 square feet ; and each boiler is normally capable of evaporating 11,000 lbs. of water per hour, with a maximum of 13,000 lbs. per hour, the working steam pressure being 160 lbs. per square inch. The grate area per boiler is 59½ square feet. The tubes are 4 in. diameter and 18 ft. long, and are arranged in four sections of ten. Each boiler is fitted with a superheater having 339 square feet of heating surface, this being sufficient to allow of the steam produced by the boiler receiving a superheat of from 100° to 120° F. After erection the boilers and superheaters were tested under an hydraulic pressure of 240 lbs. per square inch. The boiler mountings are by Dewrance.

The superheater consists of 48 solid drawn steel tubes 1½ in. diameter, and suitable piping and cocks are provided for flooding it when necessary. The guaranteed efficiency of each boiler and superheater is : 68 per cent. at maximum load, 70 per cent. at full load, 70 at three-quarter load, and 68 at half-load. About 10,800 firebricks

and 37,500 ordinary bricks were required for setting the three boilers.

The economiser consists of 240 9-ft. tubes, and is of the Claycross type. The feed water can be taken either from the river Avon or from the town mains. The scrapers have overlapping joints and chilled edges, and are driven by an electric motor.

There are two feed pumps, one steam and one electrically driven. The former is by Messrs. J. P. Hall and Son, is of their vertical slow-speed type, and was supplied by Messrs. S. Sugden, Ltd. ; the latter is a Blake and Knowles three-throw pump, and is driven by a Westinghouse direct-current motor. The Hall pump is capable of delivering from 30 to 45 gallons of water per minute with a steam pressure of 150 lbs. per square inch and 100° F. superheat against the boiler working pressure of 160 lbs. The guaranteed delivery per 1 lb. of steam at normal, three-quarters, and half-load is 72, 66 and 59 lbs. of water respectively. The Blake and Knowles pump has cylinders 5 in. diameter with 6 in. stroke, and can drive 2,700 gallons of water per hour against the boiler pressure when running at 40 revolutions per minute. This pump does not absorb more than 8·5 b.h.p. from the motor at full load, 7 b.h.p. at three-quarter load, and 5 b.h.p. at half-load.

There are also two make-up pumps by Blake and Knowles, each of these having a capacity of 300 gallons per hour.

The two surface condensers by the Wheeler Condenser and Engineering Company, Ltd., have a combined cooling surface of 2,820 square feet, and are together capable of dealing with 28,000 lbs. of exhaust steam per hour at normal load, and 35,000 lbs. for one hour as a maximum, the circulating water being obtained from the adjacent river Avon. The brass tubes are ¾ in. diameter, 18 B.W.G. thick, and seamless drawn and tinned inside and out. Two Wheeler combined air and circulating pumps were supplied by Messrs Fraser and Chalmers, Ltd. The steam cylinder is 12 in. diameter, and the air and circulating cylinders 14 in. diameter, the stroke being 12 in. The condensed water cylinder is 3½ in. diameter and the stroke 12 in. The whole of the piping for the steam exhaust, feed and circulating water was supplied by the Sir Hiram Maxim Electrical and Engineering Company.

A Holly steam-loop and gravity system is installed for taking care of the high-pressure drains, this being entirely automatic in its action, and free from all moving parts. In this system the water of condensation in the steam pipes is carried by gravity into a small receiver 5 ft. long and 12 in. diameter, placed at the lowest point in the piping system ; from which it is forced by means of a steam jet into a receiving chamber situated in the roof of the boiler-house. From this chamber it returns by hydrostatic head into the boilers, the water returning practically all its heat thereto. The system is continuous in operation, and there are no moving parts. The water softener and purifier are by Masson, Scott and Co. The former has a capacity of 300 gallons of water per hour, and the purifier a capacity of 1,250 gallons per hour.

The major portion of the oil is first removed by the Holden and Brooke extractor mentioned below, and the

Masson and Scott purifier then removes the emulsified oil as well as all oil remaining in suspension, and produces a thoroughly pure water for boiler feed purposes.

The main steam piping is 12 in. diameter, and the branches 6 in., with 3-in. pipes to the pumps. The system is fitted with Hopkinson valves, and was supplied and fixed by the Sir Hiram Maxim Electrical and Engineering Company.

The Holden and Brooke grease extractor is capable of dealing with 1,500 gallons of condensed water per hour and is fitted in the main exhaust pipe, and the latter then passes on to the two surface condensers. An automatic relief valve is fitted, and for the purpose of exhausting to atmosphere, a galvanised iron pipe, 14 in. diameter, leads up through the boiler-house roof. The main exhaust piping is 16 in. diameter, with 11 in. diameter branches to the engines.

## COAL-CONVEYING PLANT.

The coal is carried straight from the railway trucks to the storage bins by "screw" conveyors made by the Conveyor and Elevator Company of Accrington. The bins have a capacity of 50 tons. A spiral coal-conveyor brings the coal from the storage to the front of each boiler fire-grate. This can deal with 10 tons per hour when running at 30 revolutions. The trough is 14 in. diameter and $3\frac{1}{16}$ in. thick, with coal outlets 9 in. square; the spindle being $2\frac{1}{2}$ in. diameter, and driven by a Westinghouse motor.

## ELECTRIC CRANE.

The Westinghouse Company also supplied a three-motor electric crane, and erected the same in the engine-room. It is a Stothert and Pitt 15-ton overhead electric traveller with a 50-ft. span, and traverses the entire length of the engine-room.

## THE ENGINES.

The three Yates and Thom horizontal, tandem, single-crank, compound, condensing engines have cylinders 15 in. by 30 in. diameter and 36-in. stroke, and Corliss valves. They are of 320 h.p. each, and are capable of driving the 200 kilowatt generators continuously at full load with 160 lb. of steam; at 25 per cent. overload for half an hour; and at 50 per cent. overload momentarily. The engines give this output under both condensing and non-condensing conditions; the efficiency of the generators being taken as 93 per cent. The Corliss valves are worked by trip gear of the Dobson pattern, and the centre-weight governor with cross-arm is such as to keep the speed constant within 2 per cent. A sensitive knock-off appliance is fitted to the governor, the purpose of which is to disconnect the trippers and hold them in such a position as to cut off the admission of steam to the cylinder. This action takes place should the engine from any untoward cause attain an excessive speed, or should an accident happen to the governing gear.

The crank shaft of each engine is of mild forged Siemens steel, and is 16 in. diameter at the flywheel and 14 in. diameter at the bearings. The flywheel of each is 16 ft. in diameter, and weighs 18 tons. Under condensing conditions, and with the steam at 100° F. superheat, the steam consumption at full, three-quarters, and half-load does not exceed 15⅛, 15⅞, and 17 lb. per horse-power-hour respectively.

## GENERATING PLANT.

The three 200-kilowatt Westinghouse direct-current traction generators are eight-pole machines, and develop a pressure of 500-550 volts; the armatures being pressed on to the engine shafts. The armatures are of the slotted drum type, with two circuit windings so arranged that the circuit will not become unbalanced by a displacement of $\frac{1}{16}$ in. from the geometric centre of the fields. The windings are arranged to give 500 volts at no load and to over-compound to 550 volts, with a full load of 365 amperes. The insulation of both armature and field-coils is flashed with an alternating pressure of 3,500 volts.

The generators are capable of standing an overload of 25 per cent. for half an hour; and one of 50 per cent. for short periods. With a full load run of one hour the temperature rise did not exceed 40° C. above the surrounding air; while with half an hour's run at 25 per cent. overload, the rise did not exceed 50°C. At full, three-quarters, half, and quarter-load the efficiency works out at 93, 92·5, 91·5, and 88 per cent. respectively. Each of the machines was submitted to the following special test: With the no-load voltage adjusted to 500 a load of 547 amperes (with such increase of voltage as was maintained owing to the combined series and shunt excitation) was put on to the machine for a short time; the main circuit was then suddenly opened, but the machine did not "buck," neither did any serious sparking occur.

The 75 kilowatt lighting and auxiliary traction set comprises a Westinghouse high-speed vertical, enclosed, two-cylinder, single-acting, compound engine, and a Westinghouse traction generator. The high and low-pressure cylinders of the engine are 11 in. and 19 in. diameter respectively, and the stroke 11 in. The generator is a four-pole 75-kilowatt D.C. machine, developing some 150 amperes at from 500 to 550 volts, the set running at 300 revolutions per minute.

The two 15-kilowatt Westinghouse boosters are 4-pole machines running at 575 revolutions per minute, the motor volts being 500 and the booster volts 50.

## MAIN SWITCHBOARD.

The main switchboard is of the standard British Westinghouse traction type, with white marble panels two inches thick set in an angle-iron frame. There are thirteen panels altogether, the purpose of these being as follows: One each for the three large traction generators; one for the lighting set; one station-load panel; one B. of T. panel; one each for the two booster sets; one for the station lighting panel; one for the yard and car-shed lighting; two for positive feeders; and one for the negative main.

The various instruments are of the British Westinghouse types, and include a combined bus-bar and paralleling voltmeter mounted on a swinging bracket.

## THE TRACK AND FEEDER CABLES.

Mr. Charles Chadwell, of 20, Victoria Street, S.W., undertook the laying and equipment of the track and the feeder cables; the rails, fish-plates, bolts, and nuts, being supplied by the North Eastern Steel Company, of Middlesbrough.

The rails are cross-bonded every 33 yards, the bonds being "Crown" 4/0 B. and S. gauge with ⅞-in. nipples. These were supplied by the American Steel and Wire Company.

The points are 8 ft. 6 in. and 12 ft. long, and are chiefly of Hadfield's Patent Manganese steel, some of them being supplied by the Lorain Steel Company. The crossings were supplied, and the special track-work done, partly by Hadfields and partly by the Lorain Steel Company.

### OVERHEAD EQUIPMENT.

The overhead line is mostly supported on brackets (varying from 6 ft. to 22 ft. in length) on side poles ; but in some parts of the city span wires with supporting rosettes are stretched from the fronts of the houses. The latter were supplied by Messrs. Stuart and Lloyd, and the pole bases by Messrs. Wm. Bain and Co. The poles are in three pieces, and stand 24 ft. high over all. There is no centre-pole construction. The trolley wire was supplied by Messrs. Fredk. Smith and Co., Manchester, and is of hard-drawn copper 2/0 B. and S. gauge (·364 in. diameter), this being practically equivalent to 3/0 S.W.G. size.

The guard wires are of galvanised steel in strands. 7/12, 7/14 and 7/15 ; the manufacturers bein Messrs. Richard Johnson and Nephew. These are earthed through the poles, every third one of which (in the guard wire sections) is bonded to the rails.

The cars are of the ordinary type, and both mechanical and trailing frogs are used. The cars were obtained from the Electric Tramway Equipment Company, who also supplied some of the frogs, the remainder of the latter being manufactured by Messrs. Brecknell, Munro and Rogers. The average length of the sections is half-a-mile, some of the section boxes containing feeder pillars, and one of them a Board of Trade panel in accordance with the regulations. These boxes were supplied by the British Thomson-Houston Company, Ltd.

The distributing system comprises some seven and three-quarter miles of paper-insulated and lead-covered cables by the British Insulated and Helsby Cables, Ltd. They are laid partly solid and partly in porcelain ducts. There are 270 yards of ·2 square inch ; 4,700 yards of ·25 square inch, and 6,700 yards of ·3 square inch section. The return cables have a section of ·6 square inch, and total up to 1,760 yards. There are over 6,700 yards of 3-core, 4-core, 6-core, and 7-core 7/22 pilot wires ; and some 8¼ miles of overhead telephone and test wires, all this work being done by Mr. Charles Chadwell.

### THE CARS.

Of the thirty cars that are so far running, 26 are double-deckers and 4 single-deck combination cars, all of these being of Milnes construction. There is also one watering car. The double-deckers measure 27 ft. in length over collision fenders, and the single-deckers 28 ft. The trucks are of Milnes S.B. 60 4-wheel type, the wheel base being 6 ft. ; the wheels and axles being supplied by the British Griffin Company. The stairways are of the ordinary pattern, and the Milnes type of life guard is fitted to each car. The trolleys have swivel heads with graphite brushes, and were made by Messrs.

Brecknell, Munro and Rogers. The destination signs are of the British Electric Car Company's pattern, and were supplied by Messrs. Milnes. The cars are nicely upholstered, and are lighted both inside and outside. The British Westinghouse Company carried out the electrical equipment of the cars ; each being fitted with two 49 B-type 30 h.p. motors, and No. 90 controllers, as well as with the British Westinghouse Patent Magnetic Brake. There are also hand brakes.

The severe gradients and curves on the Bath system (the maximum gradient being 1 in 11) have put the Westinghouse (Newell) magnetic brake, with which all the cars are fitted, to a severe test. This brake, by the way, is the only one which has received the approval of the Board of Trade, one of its leading features being that it will, with the controller handle in the breaking position, automatically bring the car to rest on the steepest incline should the trolley leave the wire.

### THE LUNCHEON.

At the luncheon Sir Vincent Caillard was supported on the right by Lady Sivewright, the Mayor (Major C. H. Simpson), Lady Caillard, Mr. Henry Milton, Mrs. Hugh Clutterbuck, Mr. G. Woodiwiss, D.L., Mr. Swinton, Mr. Silcock, Mr. Donald Maclean, Mr. Hopkins, the Town Clerk, Mr. W. Pitt, and Mr. Crisp, and on the left by Lady Fairbairn, Sir James Sivewright, Mrs. Caillard, the Rector of Bath, Col. H. F. Clutterbuck, Mr. Lukach, Alderman Phillips, Mr. Trenow, Mr. W. S. Brymer, Mr. Huxtable, Mr. Harper, Mr. Oliver Shiras, and Mr. McCarter, while Aldermen Rubie, Taylor, Moger, Chaffin, and Stone presided at the other tables. There was a large and influential gathering.

The Mayor, proposing the toast of ' Prosperity to the Bath Electric Tramways, Ltd.,' claimed that wherever electric trams had gone prosperity had followed. It was satisfactory to know that the Bath Company had had as good, if not a better start than similar tramways in other parts of the kingdom. The trams, while they would not inconvenience carriage folk to the extent many supposed, would further extend the frontier—as it might be called—of the city, and if ratepayers were taken away it was certain they would be brought back to spend money in the city. With regard to the power station, it would be wrong for him to sit down without expressing the appreciation of Bath and its citizens on the way in which it had been constructed. In the shortest possible space of time a gigantic work had been done, and those who inspected it that day could not but have been astonished at what had been done there. To Mr. McCarter (applause) and those who had worked with him they accorded their best thanks (applause).

Sir Vincent Caillard responded, and in the course of an excellent speech recognised the cordial, energetic, and efficient assistance which, from the very outset, they had received from Sir James Sivewright. While on this topic he would like to add, for his co-directors, that he thought they had as efficient and as small a Board as any tramway company in the kingdom. Efficiency was a great thing and be believed it was well promoted by smallness. (Hear, hear.) He would also like to take that first public opportunity of recognising the most excellent solid work which their engineers—Messrs. Hopkins and Harper—had put into that undertaking, and he could not pass away from that part of the undertaking without mentioning the contractors—the Westinghouse Company—whose magnificent work at the power station they had seen, and Mr. Chadwell, who was responsible for the overhead and the roadwork. After a humorous reference to certain little tiffs with the Corporation, which had, he thought, served to draw them still closer together, he remarked that the exceedingly good start to which the Mayor had drawn attention had been made under difficult circumstances. Their power house was not ready ; they had not even now got the number of cars they had ordered, and therefore the possibilities of traffic and the convenience of the public had not been developed nearly so far as they would be. The future which was assured to the tramways was evidently so prosperous that they might almost take the good wishes they had extended in drinking that toast, as fulfilled. It only remained for him to thank the whole of the public of Bath for the fine welcome they had given to the tramways, as had been proved by the use they had made of them, and to thank the company for the way they had received the toast.

Sir James Sivewright, in felicitous terms, gave "Prosperity to the City of Bath." Referring incidentally to the discovery of radium in the deposits of the Bath mineral waters. He was glad that the responsibilities of chairman had been taken over by a gentleman so well qualified to occupy the position, and one so well known in the West as Sir Vincent Caillard. (Applause.) He knew of no tramway company with a stronger Board. There was Mr. Hugh Clutterbuck (applause), whose life had been made a burden to him by the citizens during the installing of the system, and Mr. Swinburne, a man whose name was a household word in matters connected with electricity. He was also glad to find the citizens of Bath had recognised already the sterling worth of their manager, Mr. R. D. McCarter. Their engineers had done excellent work, but after all it was to the general manager that the public looked, as well as the shareholders, for in such a public undertaking it was right that every possible consideration should be, as far as possible, shown to the public, consistently with the interests of the undertaking. There was also Mr. Trenow, secretary of the company, who, in deference to the wishes of the Board, had joined the Board as managing director. In drinking "Prosperity to the City of Bath," they who were connected with the company recognised that they were drinking to the company's prosperity also. (Hear, hear.) He wanted them to bear in mind that the prosperity of Bath and the prosperity of the trams were linked up now for better or worse, so long as the company was in existence. Whatever was to the advantage of the one was to the advantage of the other, and as one who had a large stake in the company, and who meant to retain it, remembering that they never looked for bloated dividends, but only to enjoy a fair rate of interest, he had the utmost confidence in giving "Prosperity to the City of Bath." (Applause.)

The Mayor, in the course of a brief reply, referred to the important work which had been done by Alderman Taylor, on behalf of the Corporation, in bringing the negotiations to a satisfactory termination.

A few remarks from Alderman Taylor followed. He believed the trams had come there not only to stay but also to pay. As far as he and his committee were concerned, they were perfectly satisfied with the transactions they had had with the company, and their friend, Sir James Sivewright. (Applause.) The system was quick, cheap and safe. (Applause.) That the trams were fully appreciated was shown by the fact that during a single week they had carried double the population of the City. (Applause.)

### CONGRATULATIONS.

In conclusion, we may congratulate the engineers (Messrs. George Hopkins and Son, Messrs. Harper Bros. and Co., and Mr. R. D. McCarter) and the contractors (Mr. Charles Chadwell and the British Westinghouse Electric and Manufacturing Company) upon the quick completion of their work ; great credit being also due to Mr. R. D. McCarter for the efficient way in which he superintended the erection and equipment of the generating station, etc., and also for his successful inauguration of the duties of management.

# PAGE'S MAGAZINE

An Illustrated Technical Monthly, dealing with the Engineering, Electrical, Shipbuilding, Iron and Steel Mining and Allied Industries.

## DAVIDGE PAGE, Editor,

Clun House, Surrey Street, Strand, London, W.C.

Telephone No : 3349 GERRARD.

Telegraphic and Cable Address : "SINEWY, LONDON."

**Editorial.**—*All communications intended for publication should be written on one side of the paper only, and addressed to " The Editor."*

*Any contributions offered, as likely to interest either home or foreign readers, dealing with the industries covered by the Magazine, should be accompanied by stamped and addressed envelope for the return of the MSS. if rejected. When payment is desired this fact should be stated, and the full name and address of the writer should appear on the MSS.*

*The copyright of any article appearing is vested in the proprietors of* PAGE'S MAGAZINE *in the absence of any written agreement to the contrary.*

**Correspondence** *is invited from any person upon subjects of interest to the engineering community. In all cases this must be accompanied by full name and address of the writer, not necessarily for publication, but as a proof of good faith. No notice whatever can be taken of anonymous communications.*

*The Editor does not hold himself responsible for the opinions expressed by individual contributors, nor does he necessarily identify himself with their views.*

### Subscription Rates per Year.

**Great Britain**– In advance, 12s. for twelve months, post free. Sample Copies, 1s. 4d., post free.

**Foreign and Colonial Subscriptions, 16s.** for twelve months, post free. Sample Copies, 1s. 6d. post free.

Remittances should be made payable to PAGE'S MAGAZINE, and may be forwarded by Cheque, Money Order, Draft, Post Office Order, or Registered Letter. Cheques should be crossed "LONDON & COUNTY BANK, Covent Garden Branch." P.O.'s and P.O.O.'s to be made payable at East Strand Post Office, London, W.C. When a change of address is notified, both the new and old addresses should be given. All orders must be accompanied by remittance, and no subscription will be continued after expiration, unless by special arrangement. Subscribers are requested to give information of any irregularity in receiving the Magazine.

### Advertising Rates.

All inquiries regarding Advertisements should be directed to " THE ADVERTISEMENT MANAGER, Clun House, Surrey Street, Strand, London, W.C."

### Copy for Advertisements

should be forwarded on or before the 3rd of each month preceding date of publication.

# OUR MONTHLY SUMMARY.

LONDON, February 22nd, 1904.

### The War.

We have heard enthusiasts express longing for a great naval European war in order that the strength or weakness of specific battleship construction might be demonstrated. The historic engagements of the past month have deprived them of all excuse, and at the same time have furnished our naval experts with some valuable data. Whatever may be our opinions as to the *casus belli*, every Englishman must feel a sense of satisfaction in knowing that the Japanese navy is almost entirely of British construction. It seems scarcely credible that prior to 1859 the Japanese fleet consisted of a number of junks and one or two ships built from Dutch designs of the seventeenth century. Japan's splendid navy is a standing monument to her enterprise, and her audacity in attack shows that the personnel of a modern navy is a factor of no less importance than it was in the days of Queen Elizabeth. The war is specially dealt with in this number by our naval Correspondent.

### Problems for the Engineer.]

In the course of his presidential address to the Manchester Association of Engineers, Mr. Alfred Saxon dealt with the forces that have gone to make up what he termed an engineering revolution. The conclusion of the address was concerned with the future of the mechanical engineer and the problems he will have to face " On what lines," he asked, " might they expect this engineering revolution to continue ?" The demand for producing electricity cheaply for power and lighting ; the demand for the cheapening of manufactured goods and articles of all descriptions ; the shortening of the time occupied in ocean voyages ; the development of aerial navigation ; the demand for carriage of goods and merchandise at a lower cost, and amongst the problems of our city life the demand for cheap and rapid transit to and from the suburbs. They had also to consider the problem of the reduction in the losses which existed in power installations, and the reduction of the inertia and frictional losses in all kinds of machinery and machine tools, the existence of which to so considerable an extent had been revealed by electric driving. There was also the reduction of losses occurring in the transmission of power, and generally their aim should be to increase the amount of useful work done in proportion to the total power expended. New sources of power were being introduced from time to time, and practically tested ; other sources were being suggested, and so active was recent mechanical and chemical research that they had amongst other discoveries the marvels of radium to add to the list. They had inferior coals, oil, peat, vegetable mud, natural gases, and the utilisation of other substances which had hitherto been looked upon as waste materials to augment their coal supplies, and there was no doubt that they have other natural resources hidden which had not yet been tapped, and which would repay their greatest efforts to discover. The future of engineering was bright with possibilities, and there were those who saw in means of increased production and improved facilities of transport a time when the working days would be shortened, but if these dreams were to be realised, engineers must not in their work choose the path of least resistance, but must tackle in their day and generation the problems that arose. The country was no doubt at a disadvantage with many other countries in the generation of electricity in the cheapest manner by water turbines. Therefore from a national point of view, economies were absolutely necessary if they were to conserve that form of energy contained in coal supplies, which had hitherto helped them in the industrial race.

## Employment in 1903.

The Board of Trade returns regarding employment last year do not make very cheerful reading. It requires no extreme of optimism to predict a better state of things next year, for 1903 showed a falling off compared with the three years immediately preceding and was not up to the level of an "average year." The mean percentage of unemployed returned by Trade Unions during 1903 was 5·1, compared with 4·4 in 1902, 3·8 in 1901, 2·9 in 1900 (a year of exceptionally active employment), and 2·4 in 1899. The average percentage for the ten years 1894 to 1903, was 4·1. The falling off in 1903, as compared with 1902, was marked in the latter half of the year. There has been a continuation of the decline in demand for labour in the building trades which has been going on since 1900. Shipbuilding was acutely depressed in 1903 by a lessened demand for new ships (connected with the low level of freights during the last three years) and at the present time, there is a high percentage of workpeople unemployed in this trade. The trades (e.g., steel) which supply the shipbuilders with materials were also affected adversely Employment for coal miners, as shown by the average weekly numbers of days worked by coal pits, was less active in 1903 than in any year since 1896.

On the other hand, employment in 1903 was fairly good in certain industries, and among these may be mentioned iron mining and tinplate manufacture.

## The British Workman and Cousin Jonathan.

While on the subject of employment, we are reminded of some interesting comparisons which have been lately made between the general conditions here and in the shops of America. For instance. "W. H." in "The Mechanical World" says that a return to England after a few years in America is very relaxing. One wonders comparatively what English workmen have to complain about. Why don't they do a reasonable amount during the few hours they are at it? Their pay is little enough, certainly, but it is not as well earned as the American's extra money. It makes one feel uneasy regarding England's future to see that the prevailing spirit is to give as little work as possible in return for wages paid. "W. H." quite understands the altruistic spirit which is largely the cause of it; but looked at from an international standpoint, it is like one-sided free trade. You don't find it in American shops to any appreciable extent. The difference cannot be accounted for on the ground that the American is paid better, and treated better. The English workman, generally speaking, is treated as he wants to be treated, and has the conditions that suit him. Conditions that are entirely against his tastes cannot be imposed on him. Personally, continues "W. H." I prefer English conditions to American—with the exception of wages, of course; but I should like to see English shops make for more freedom and equality between men on different lines from those prevailing in America, though in some respects America might well be followed.

Comfortable working conditions is a thing we pay little attention to. The American works under the best conditions he can get. In the course of further observations he remarks that there is only one thing wanting in America, and that is a superior class of men to correspond with the superior material conditions. This America has not yet got.

## The "Ideal" Employer.

Having glanced with " W. H." at the respective conditions of English and American workmen, let us devote a minute or two to the "ideal" employer from an American point of view. If the reader has a large perceptive mind, if he can see a dozen different objects at one time while engaged in conversation on another subject, and can answer questions mechanically yet correctly, he fulfils the first condition laid down as necessary for the ideal employer, by Mr. R. A. Baker, in a paper contributed to the Cincinnati Metal Trades' Association. We are rather inclined to think that such a person would prove somewhat irritating, as his fits of abstraction might not be confined to the workroom—so perhaps it is just as well that he is only ideal. The author goes on to say that he should possess great intellect, be broad in his views, both kind and just, ever ready to mete out scorn or praise, wherever the call, whether friend or foe, relative or stranger. Let the ideal employer create an incentive for better work, show the employee where he is weak, have experts go over all details to be produced, devise means, jigs and charts, and positive instructions how to produce each part as economically as possible, mechanically true and correct. Post these instructions where the men can view and study them. Stir their brains to activity and success will be your reward. The ideal employer should not permit himself to allow the shop management to become a strictly family affair or be made up of influential friends and deadheads regardless of merit and ambition. He should be a born leader of men, of sweeping influence, easy of approach, capable of condescending to a friendly word for this or that employee, without endangering their respect for him. He should very thoroughly and carefully look after the general health and welfare of his employees, and must pursue a policy along friendly and co-operative hues. Bring business methods into their work, says the author, and they will realise the trust confided in them in handling your dollars.

## Radium—Some Misconceptions.

An American contemporary reminds us that in nearly all the popular articles and many scientific ones, radium is a word used to denote not the metal of that name, but its chloride or bromide, generally the latter. In a similar slipshod fashion the photographer sometimes refers to the silver in an emulsion, meaning thereby the bromide or chloride of silver. When the statement is made that radium decomposes or gives off emanations or rays, it is wholesome to remember that the bromide of radium generally quite impure, is the substance having these properties. As to the amount and character of the impurities referred to, it would be unwise to hazard a guess. Madame Curie, like her distinguished husband, is far more cautious and conservative in statements about radium than those whose knowledge is derived merely from a few decigrams of the stuff in a sealed tube.

One of the most curious statements about Radium we have lately seen had also an American source. The discovery of radium in the deposits of the mineral waters of Bath is still fresh in mind. The news was apparently cabled over to the States, and in a certain technical journal we read the following :—

London's ancient city bath has come to the front this week because of the discovery that the old hot

# Our Monthly Summary.

baths contain, in the waters they have been throwing up for centuries, no end of radium which has gone down the throats of invalid drinkers or has been disported in by bathers who must have numbered millions. That excitement has been great over the discovery is shown by the columns which have been published in newspapers and letters discussing the wonderful find.

It all came about in this way : The Hon. R. J. Strutt, son of Lord Rayleigh, while analysing the waters of the bath, found, as he states in a letter which he sent to the Municipal Council of this city, that the waters contain radium in appreciable quantities, etc.

Such is fame ! Anyone who has handled " flimsy " will readily understand how this enterprising city of Bath came to be mixed up with London's " ancient city bath," but we cannot help wishing that some of our American friends would come over and know Bath for themselves—more especially since the city has just inaugurated electric trams, and is generally becoming as up-to-date as is consistent with her classic traditions.

It is of course true that there is a London Roman bath near the Strand, and within twenty yards of these offices, but up to the time of going to press we have not heard that any radium has been found in it.

### The Institution of Mechanical Engineers.

A very satisfactory report was that presented by Mr. Worthington at the annual general meeting of the Institution of Mechanical Engineers. The total number in all classes on the roll of the institution at the end of 1903 was 4,211, as compared with 3,892 at the end of the previous year, showing a net gain of 319 as against the gains of 238, 243, 325, and 402 respectively during the years which have elapsed since the institution took possession of its new home. Among members removed by death during the year must be mentioned Sir Frederick Bramwell, who joined the institution fifty years ago, and was president in 1874 and 1875 ; Mr. F. C. Marshall, who was elected member of Council in 1882, and Mr. George B. Lloyd, whose membership dated from 1854. A satisfactory balance sheet was presented, showing a revenue for the year of £10,848.

The report mentions with regard to the Alloys Research Committee that notwithstanding the gratuitous services of those who have organised the experiments, the expenditure upon this research since the formation of the committee in October, 1889, has exceeded £1,800. The first report of the Steam Engine Research Committee by Professor D. S. Capper has been received, and will shortly be presented at a meeting of the institution. It is interesting to note that the Reference Section of the Library (consisting of about 4,000 books and pamphlets) has been re-arranged on the Dewey decimal plan.

The spring meeting will be held this year, conjointly with the American Society of Mechanical Engineers, in Chicago, beginning May 31st, and a visit will afterwards be paid to the St. Louis Exhibition.

### Proposed Thames Dam.

The scheme to construct a dam across the Thames at Gravesend was recently considered at a well-attended meeting of wharfingers, shippers, and others interested in the Port of London. Sir Thomas Brooke Hitching, who presided, said that the scheme which they had met to consider was proposed many years ago by Mr. J. Casey, a member of the Institution of Naval Architects, and had since been practically worked out by Mr. T. W. Barber, a member of the Institution of Civil Engineers, who called it the Thames Barrage. It was felt that great advantages would accrue to wharfingers and others if the river were barred at Gravesend, as it would enable them to use their wharves all day long, instead of, as at present, only during four hours out of every twenty-four. It would also have the result of making the water as clear in the lower reaches as it was above Richmond and Teddington. The London County Council, in their proposals to establish a steamboat service on the Thames, had always been faced by great tidal difficulties, but with a barrage, giving constant high water, it would be possible to abolish the existing Thames Conservancy piers, which ran out into the river, and to have passenger steamboats as small as those which plied on the Seine and the Elbe. Moreover, the general appearance of the river would be greatly improved by what he had heard described as the canalisation of the Thames.

Mr. J. Casey pointed out that the scheme would allow large vessels to proceed without delay to the upper reaches of the Thames, and so save a large amount of the present cost of cartage.

Mr. T. W. Barber explained that the proposed dam would be of solid masonry, and would rise above the level of the highest possible tides. On the top of the wall would be a roadway, which would connect Kent and Essex, and underneath there would be a tunnel, which could be used by the railways north and south of the Thames. In the centre of the dam a number of locks would be made, capable of taking the largest ships afloat. In addition to the locks, there would be weirs over which the surplus water would flow away into the sea. Gravesend was considered the best part of the river at which to construct the dam. The dredging proposed by the London County Council and in the Government Bill could not be carried out. He maintained that the river could not be dredged to a depth of 15 ft. below its present bottom ; even if it could be done, the river walls would fall in. It was important that this barrage scheme should be threshed out before a Parliamentary Committee, as it was a question of spending only £3,000,000 to £4,000,000, as compared with the £40,000,000 with which the ratepayers would be saddled under the Port of London Bill.

# THE CIVIL ENGINEER AT WORK.

## By C. H.

### Railway Developments in China.

Mr. Arthur Judson Brown, in the course of an article contributed to the American Review of Reviews, recalls the fact that the conservatism of the Chinese for many years proved too strong for the promoters of steam railways. In fact the first line, which covered the fourteen miles between Shanghai and Wu Sung was no sooner completed by its British promoters than the Government bought it, tore up the road bed and dumped the engines in the river. This was in 1876. Since then the Chinese have had many object-lessons. If the vast schemes which are at present contemplated by companies of varying nationality can be realised, there will not only be numerous lines running from the coast into the interior, but a great trunk line from Canton through the very heart of the empire to Peking, where other roads can be taken to Manchuria and Korea, or to any part of Europe. The far-reaching effect of this extension of modern railways will, of course, mean a new era for China. The author, after pointing out her agricultural possibilities and immense deposits of coal and iron, remarks that to make these resources available to the rest of the world and in turn to introduce among the

426,000,000 of the Chinese the products and inventions of Europe and America will be to bring about an economic transformation of stupendous proportions.

A curious difficulty but a very formidable one encountered by the railway engineer in China is the omnipresence of sacred graves. This is owing to the peculiar Chinese custom of burying their dead wherever a geomancer indicates a "lucky" place.

### Dredging in Montreal Harbour.

Some very interesting details of the dredging fleet of Montreal Harbour were recently presented to the Canadian Society of Civil Engineers by Mr. H. A. Bayfield, A.M.Can.Soc.C.E.

The Harbour Commissioners' dredging fleet proper at present consists of four dipper dredges, five floating derricks, one drill boat, five tugs, and twenty-three scows. There are other machines, but as the work done by them cannot be said to pertain to dredging, no particular mention is made of them. It may be stated, however, that the complete fleet comprises forty-one vessels.

The dredges are all of the dipper or spoon type, their principal dimensions being as follows : Length of hull

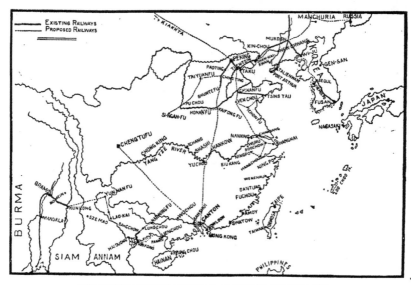

MAP SHOWING CHINESE RAILROADS, COMPLETED AND PROJECTED

90 ft., width of hull 36 ft., maximum depth of hold from 9 ft. 6 in. to 10 ft. 9 in., according to dredge ; size of main engine 16 in. by 18 in. double cylinder.

The standard bucket employed is of seven cubic yards capacity, and is built with a curved lip and front. There are also a few straight-lipped dippers of five and a quarter yards capacity, but though specially designed for dredging very hard material they have proved not at all superior to the larger bucket , and are, therefore, used only for work in a strong current where swinging is difficult with the dipper of large size, or when filling boxes which the seven yard bucket would overload. The body or shell of both sizes of dippers is of steel plate, with butt joints and single cover strap. The lip and lower band are of cast steel and are riveted to the shell, riveting being countersunk on the inside to insure a smooth surface. The door is composed of a single piece of inch and a quarter steel plate riveted to a cast steel hinge piece. The latching device is on the toggle joint principle, and, considering the extremely severe usage it is subjected to, gives good satisfaction. Bails are a single steel casting, secured to the hoisting wire socket a cast steel shackle. The shackle takes up considerable room, but its use is necessary to prevent injury to the wire by sharp bending at the point where it enters the socket.

Each dipper carries four teeth which are of cast steel with chisel points. The teeth are made hook-shaped, the hook fitting over the lip at places where projecting lugs are provided to prevent side play. The lower end of each tooth is secured to the shell by four bolts. With this arrangement of fastening, a change of teeth can be effected in about half an hour.

### Experiments and Improvements.

Several attempts have been made to design a dipper tooth that will last a reasonable length of time in hard digging. Detachable points of hardened steel were experimented with, but proved a failure through lack of strength either in the point itself, or in the body of the tooth which had to be cut away to receive it.

Silver tips were V welded into the teeth and given a trial in rock digging. It was found that when the points were tempered sufficiently to wear well, they were brittle and broke too frequently to warrant their use.

At present the only effort being made to improve the lasting qualities is confined to heating and dipping the tooth point after it is drawn out under the hammer.

The life of a set of teeth will of course depend upon the nature of the material being dredged. In unblasted grey rock seven hours continuous work is a fair average, as shown by records kept by the writer. Upon becoming too blunt to hold well, the points are again hammered out, but there is a limit to the drawing out process, and the tooth soon goes to the scrap pile.

During last season the four dredges of the fleet used up 181 teeth, each tooth having been in service at least four times.

Considerable trouble has been experienced on account of the very rapid wear of the faces of the latching dogs, even though they have until very recently been made of Manganese steel. Castings of common steel with faces hardened as much as possible are now being tried, and though the duration of the test will not allow of a decision as to the wearing qualities, it is certain that they withstand shocks and blows far better than those of Manganese steel.

Experiments are in progress to determine the advisability of annealing all steel castings used in the construction of buckets and clams. There is no doubt that the initial strain in some of these castings is very high, as its presence is frequently denoted by shrinkage cracks. It is hoped that annealing will relieve more or less of this tension and materially increase the life of the piece.

With regard to riveting, it has been proved that rivets driven by pneumatic tools are superior to those put in by hand. The drift pin seems unavoidable in work of this kind, but it is certain that were all holes drilled fair and good, slack rivets and broken castings would not be so common.

### A Model Street.

The new Kingsway from Holborn to the Strand will be a model thoroughfare in many ways, but more particularly by reason of its complete immunity from those periodic upheavals which are characteristic of some of the adjoining streets. On the occasion of a recent visit, the members of the Civil and Mechanical Engineers' Society showed great interest in the shallow tramway works under the new artery, and the difficulties that have been encountered in connection therewith. The subway, starting from a spot close to the New Gaiety Theatre, will pass under the western arm of Aldwych and then north under Kingsway to Holborn. It will then pass under Holborn to Southampton Row, where it will come to the surface at a slope so as to join up with the existing tramway lines that concentrate at Theobald's Road. On each side of the tramway tunnel there will be two subways, which will accommodate gas and water pipes, electric mains, etc. Below each of these side tunnels there will be a sewer. The visitors were first taken into the short length of tunnel at the Strand end, which is nearly completed. Near here, close to the site of the old Olympic Theatre, there will be a station, which will be reached by staircases, but the whole scheme includes a continuation of the tunnel as far as the Embankment. The width of the tunnel will be 20 ft., and the height from the crown of the arch to rail level 14 ft. The pipe subways on each side of the main tunnel will be 12 ft., whilst the egg-shaped sewers beneath will be 2 ft. 8 in. by 4 ft. 6 in. The depth of the tunnel beneath the surface varies from 6 ft. to 14 ft., the greater depth being where the work passes under Holborn and the Strand. At these places the main tunnel will be divided into two smaller tunnels for convenience of construction. Special precautions have been taken to preserve the brickwork from decay by damp, there being a lining of asphalte three-quarters of an inch or more thick. The electric tramcars will be supplied by current from conductors laid in conduits.

# ELECTRICAL AFFAIRS.

BY

## E. KILBURN SCOTT, M.I.E.E., A.M.Inst.C.E.

### Separating Non-magnetic Metals by Electricity.

There appears to be a general impression that it is only possible to separate metals by electrical means when those metals are magnetic. The usual plan is to pass the metal in a finely-divided form in front of a series of electro-magnets. It is probable, however, that any metals, whether they are magnetic or not, may be separated by moving magnets in front of the materials, so that currents are induced in the mineral particles. No one can move copper, for example, merely by approaching it with a magnet, but if a series of magnets of differing polarities pass it in quick succession, then the copper will tend to move by reason of the currents induced in it. In other words, when a rapidly alternating current is generated in a piece of metal, that metal would rather pass off in the direction of the magnet than have electrical currents flow through it. Sir Hiram Maxim appears to have been an early worker in this field, and it is one in which there would appear to be considerable scope.

### A new Alternating Current Flame Arc Lamp.

Messrs. Oliver and Co. have developed an alternating current " flame " arc lamp, which is on an entirely different principle to any hitherto set to work. A few particulars may therefore be of interest.

Two flame carbons, each 8 millimeters in diameter and 18 in. long, are placed side by side in metal tubes, the space between being about ¾ in. Alongside the carbons there is a glass rod about one-sixteenth of an inch in diameter, and the same length, in a separate metal tube. The arc is struck by means of a piece of carbon mounted on a short bell-crank lever, which rests against the ends of the two flame carbons. Immediately current is switched on, a solenoid pulls the short circuiting piece away, so striking the arc. The end of the glass rod projects very slightly from the metal tube, and its upper end carries a weight which also rests on the top of the two flame carbons. By an ingenious magnetic deflecting device the arc plays on the lower end of the glass rod and in due time melts off a small piece, thus allowing the rod to drop about one-sixteenth of an inch. The flame carbons are then pushed down the same amount. It should be mentioned that the carbons are held up by weighted friction clutches. The two carbons and the glass rod are thus burnt away at the same time. The feature of the lamp is that there are no solenoids other than the one for working the bell-crank lever, the arc is also horizontal, and therefore gives no shadow from the lower carbon. One of these arc lamps on the Leeds Market circuit takes 34 volts 10 amperes, and absorbs 400 watts on the primary side of the step-down transformer.

### Winding Field and Transformer Coils.

In winding copper wire on to a field coil, it is usual to place a brake on the spindle of the spool which is being unwound so as to keep the wire taut. This naturally gives a side pull on the bearings of the lathe, and there is considerable friction. It occurs to the writer that if two spools were wound on the same lathe and the wires guided on to the spools from opposite directions, say top and bottom, these side pulls would be neutralised to a large extent.

The present tap tapping of the wire with wooden mallets is anything but satisfactory, and what appears to be wanted is a change gear, which would move a small pair of gripping wheels, the exact distance represented by the diameter of the wire, at each revolution. A knock-off reversing motion would, of course, be necessary to start the wire back again when it reached the end of each layer.

It almost goes without saying that there should be a revolution counter on every winding lathe, otherwise it is difficult to ensure that the exact number of turns will be wound on. With the shunt coils of a multipolar dynamo it is most important that each pole should give the same number of ampere turns, otherwise there will be unbalancing. As the coils are connected in series, the number of turns must be the same on each coil.

In the case of similar sized transformers which have to bank together, it is also important to have the turns exactly the same on both primary and secondary coils.

### Steel Conduit Tubing.

The use of steel conduit tubing for carrying electric cables is very rapidly taking the place of wooden casing. Among the debatable points in connection with the use of such tubing is the question whether the interior should be lined. Soapstone powder is used to assist the drawing in, but by lining with some smooth material the cables can be drawn in without any risk of abraiding to the insulation. At the same time, such material may act as an insulator. In one particular make of conduit the interior is lined with one-sixteenth of papier mache, impregnated with bitumen.

Another point in connection with steel tubing is the question of sweating on the interior surfaces, this being due to the fact that the temperature inside is raised by the current flowing through the cable. There is much less risk of such sweating occurring if the tube is lined, but in any case it is probable that it would be of considerable benefit to withdraw the moist air from the interior of the conduit after it has been installed, in much the same way as is done with telephone wires.

The few accidents which have occurred with steel tubing have been due to the fact that the tubing was not properly earthed. This is a most important matter, and it is not only necessary to see that the tubing is well earthed, but also to be sure that there is good conductivity between the joints. Such joints should be metal and any white lead or paint which is employed for making the conduit airtight should be beyond the screwed portion.

# POWER STATION NOTES.

## By E. K. S.

## The Drainage of Steam Pipes.

In most steam-pipe systems working with ordinary saturated steam, it will be found that there is a stream of vapour moisture flowing along the bottom of the piping. (Tests for wetness of steam should therefore be made at the bottom of the pipe, and not at the centre.) It happens that this moisture is not only a danger when it manages to get into the engine cylinders, but it has also the effect of washing out the cylinder oil.

In this country the usual plan is to provide numerous steam-traps to catch this water, but in the States, a steam-loop system such as the Holly is generally employed. This system has been fitted to the Bath Tramways' power-house by the Westinghouse Co., so a short description may be of interest. In its simplest form, the Holly system consists of a receiver placed below the lowest point to be drained, a riser pipe, and a drop leg, the whole forming a loop to the main steam range.

The riser pipe and drop leg are carried high enough to enable any water which accumulates in the drop leg to force its way into the boiler by gravity. The reason for the water and moisture finding its way up the riser pipe is because means are provided (by condensation) for causing a lowering of pressure at the top of the pipe. Naturally the steam tends to flow towards this reduced pressure point, and in so doing it carries with it any primage or condensation water and vapoury moisture. Suppose the pressure in a given boiler is 100 lb., the pressure at the engine 95 lb., and in the drop leg 94 lb. There will then be 6 lb. difference between the boiler and the drop leg, and a column of water in this drop leg standing 14 ft. above the level of the water in the boiler will provide sufficient hydrostatic head to balance the difference in pressure.

By maintaining the necessary difference in pressure at the high point of the system, the drop leg column reaches a height corresponding to this constant difference and rises no further. It is then in full action, and maintains circulation so long as steam is in the system.

The method has the advantage of being absolutely automatic and comprehensive in action, and it operates whether much or little water is present and at all times, day and night, without any attention whatever. As the water returned to the boilers is absolutely pure, it is possible to reduce scale by concentrating the water into one or two boilers at a time. In one case, the system provided the entire feed in one boiler in a battery of twelve.

## The Tangential Water-Wheel.

The simplest of all the prime movers which are in use to-day is undoubtedly the Pelton wheel, or, as it should more properly be called, the tangential water-wheel. It is closely associated with the development of electric transmission in the Western States of America, where it has been developed from the old Hurdy Gurdy wheel of the gold miners. In the modern wheel the water strikes double-curved buckets, and the impulse causes rotation. Further, the buckets are so shaped that the direction of the jet is reversed almost back upon itself, and this reversed flow, as it emerges, is reactive, and tends to further increase the power and speed of rotation..

As compared with the turbine, the first cost of the tangential wheel is very much less, and for high heads *i.e.*, several hundreds of feet, it can be used where a turbine is quite impossible. At the same time its

cost of maintenance is but a fraction of that of the turbine, for a complete set of new buckets costs very little, and can be mounted in an hour or so by an unskilled man. Regarding efficiency, any good make of wheel will give over 80 per cent., and there is this to be said, that it maintains its efficiency, whereas a turbine constantly decreases as the wheel wears away in the casing, thus allowing the water to pass by instead of through.

A disadvantage of the turbine is that the variation in design is limited by reason of its excessive cost for large diameters. The tangential wheel, on the other hand, permits of very wide and free scope. Single wheels have been built to give 3,000 h.p. on a 1,500 ft. fall and a speed of 240 revolutions per minute. The highest heads for which tangential wheels have been built are 2,530 ft. at Seattle, 2,250 ft. at Panuco Mine, Mexico, and 2,100 ft. at Pike's Peak, U.S.A. These give pressures of upwards of 1,000 lb. per square inch at the nozzle, so it will be seen that the engineering of these high falls is no child's play.

## Advantages of High over Low-Head Falls.

A glacial stream or mountain torrent is much better for the purpose of power work than water at a lower altitude, because when the water is led into pipes at a point above the tree-line there is so much less risk of organic matter getting past the strainer. In the case of a river on the other hand, expensive gratings and strainers are necessary to catch wood, leaves, and other smaller floating obstructions. In some power installations a gang of men is kept constantly at work clearing the strainers.

A low fall is generally in a valley, where land is valuable and where, if the water rights are interfered with there may be legal difficulties. A mountain stream, on the other hand, generally runs through land of little value. It should also be remembered that with the small amount of water required from a high head it is much easier to provide for storage to carry over periods of drought. A few feet difference in the height of such a reservoir makes little or no difference to the power, whereas on a 50 ft. fall two or three feet in height of the storage reservoir or the tail race makes a considerable difference in the power available.

## High-Head Water Powers.

Travellers, as a rule, are only impressed by falls of moderate height over which the water comes tumbling in masses. They do not appreciate the fact that a mountain stream dropping 1,000 or so in a short horizontal distance may be capable of developing much more power at a fraction of the expense. The result of this is that, with the exception of California and the Western States of America, very little has been done to develop high-head water powers.

In the Colonies, South America, and Africa, as well as India and the Far East, there are numerous high-head falls waiting to be put into use for the purpose of electric transmission. The rapid development which has taken place in the Western States, and the enormous distances to which the current is transmitted (80, 99, 142, and 180 miles are four of the longest transmissions) show that the generation of electricity from high-head falls is a paying business.

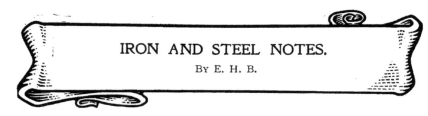

# IRON AND STEEL NOTES.

## BY E. H. B.

### Mr. R. A. Hadfield and His Work.

Mr. R. A. Hadfield has been contributing to the "Iron and Steel Metallurgist and Metallographist," published at Boston, Mass., some valuable notes on iron and steel alloys. He remarks that the examination of the properties of iron alloyed with other elements is being continued with increasing vigour, and valuable results, both from the immediate practical as well as the scientific point of view, are accruing to the world· from the labours of the metallurgist. Nowadays we do not hunt for the philosopher's stone, but verily the metallurgist has produced "transmutations" that even ten years ago would not have been thought possible. The writer has produced alloys with tensile strengths all the way from 18 tons up to 110 tons per square inch, and with elongations from nil to 70 per cent. But, singular as it may seem, notwithstanding the important part played by all these new iron alloys, carbon still maintains its premier position in determining the practical value of the various products, in other words, there are few, if any, iron alloys in which, apart from the effect produced by the special element added, the presence of carbon is also unnecessary; therefore, whatever the theory believed in by each reader of this article, that is, whether he is a carbonist or allotropist, it has to be admitted that carbon alone is the predominant factor in determining the utility of the alloy. The writer is not stating this in any controversial spirit, as, of course, the allotropist, whilst admitting the importance of carbon, claims its action is different to that believed in by the carbonist, but the fact remains that carbon must be present to render the alloy of practical value.

Finally, says Mr. Hadfield, may it not be claimed that the world is more than ever indebted to that indefatigable body of men, the metallurgists, who labour on, year by year, through difficulties often apparently insurmountable, to improve and perfect the qualities of iron for the general benefit of mankind.

As the article is for the first number of the enlarged "Metallographist" of America, Mr. Hadfield takes the opportunity of wishing God speed to the members of his craft in their work in that great Republic, where he has always received so much personal kindness and courtesy. He offers every good wish for continued success to the editors in the continuation of their efforts to improve and enlarge the stores of information on metallurgical matters.

### Transvaal Ironworks.

The *African World* states that the site of the proposed new ironworks in the Transvaal will not be at Onderstepoorte, north of Pretoria, as was originally intended. Although the great project will be carried out in its entirety, certain bank assays on the Onderstepoorte ore, made at Johannesburg last May, have been found to be quite untrustworthy.

### The Iron and Steel Institute.

The annual general meeting of the Iron and Steel Institute will be held at the Institution of Civil Engi-

neers on Thursday and Friday, May 5th and 6th. The annual dinner will be held—under the presidency of Mr. Andrew Carnegie—in the Grand Hall of the Hotel Cecil, on Friday, May 6th.

### Meeting in the United States.

The autumn meeting will take place in New York on October 24th, 25th, and 26th. After the meeting there will be an excursion to Philadelphia, Washington, Pittsburg, Cleveland, Niagara Falls, and Buffalo, returning to New York on November 10th. During the trip night travelling will be avoided, and every endeavour will be made to obviate fatigue. The two Sundays will be spent at Washington and at Niagara Falls. The approximate cost of the stay in the United States is estimated at £25. For the convenience of members desirous of visiting the St. Louis Exhibition, arrangements will be made for a limited number to leave Pittsburg for St. Louis and Chicago, reaching New York on the evening of November 10th. This trip will necessitate three nights being spent in sleeping cars, and the approximate cost will be £35.

An influential committee has been formed in the United States for the reception of the Institute, Mr. Charles Kirchhoff being the president and Mr. Theodore Dwight (99, John-street, New York), the hon. secretary.

### The Heat Treatment of Steel.

A very interesting work on the "Hardening, Tempering, Annealing, and Forging of Steel," by Joseph V. Woodworth, is to hand from Messrs. Archibald Constable and Co., Ltd. It deals with the practical side of a subject that is now very prominent, and presents a great deal of useful information, in a compressed and easily accessible form. To indicate the style of the work I cannot do better than quote one or two paragraphs.

### Points to be Remembered.

To heat and cool steel properly, says Mr. Woodworth, remember the following : Never heat a piece of steel which is to be annealed above a bright red. Never heat a piece to be hardened above the lowest heat at which it will harden, and the larger the piece the more time required to heat it is required, which will have to be higher than a smaller piece of the same steel, because of the fact that a large piece take longer to cool than a smaller piece, as when a large piece of steel is plunged into the bath a large volume of steam arises and blows the water away from it, thus necessitating more time in the cooling. Thus when the tool or die is very large, a tank should be used to harden it in, into which a stream of cold water is kept constantly running, as otherwise the red hot steel will heat the water to such a degree that the steel will remain soft.

### Different Quenching Baths—Their Effect on Steel.

On this question the author remarks that, next to proper heating, more depends upon the quenching

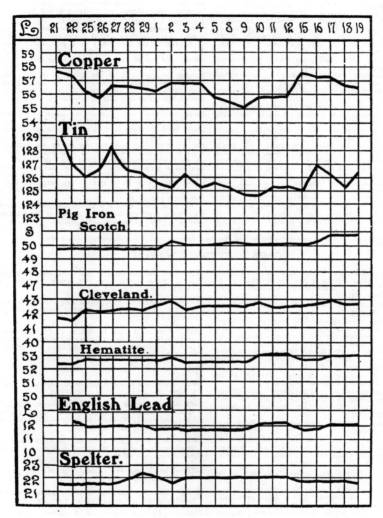

THE HOME METAL MARKET.

Chart showing daily fluctuations between January 21st and February 19th, 1904.

than anything else. It follows that the effects of the use of the various kinds of baths are required to be understood. The most generally used bath is usually cold water, though not infrequently salt is added or a strong brine is used. The following will be found to answer well for the work mentioned : For very thin and delicate parts, an oil bath should be used for quenching. For small parts which are required to be very hard, a solution composed of about a pound of citric acid crystals dissolved in a gallon of water will do. For hardening springs, sperm oil , and for cutting tools, raw linseed oil will prove excellent.

Boiled water has often proved the only bath to give good results in a large variety of work, the parts requiring hardening being heated in a closed box or tube to a low red heat and then quenched. Sometimes the water should be boiling, at others quite hot, and then- again lukewarm. Experience will teach the operator which is the best for special work. If a cutting tool. such as a hollow mill, a spring threading die, or a similar tool, is to be hardened in a bath of this sort, dip it with the hole up or the steam will prevent the liquid from entering the hole and leave the walls soft. A tendency to crack will also prevail if this is not done. The generation of steam must be considered when hardening work with holes or depressions in it, and attention must be paid to the dipping of the part so as to prevent the steam from crowding

the water away. Clean water steams rapidly, while brine and the different acid solutions do not.

### Hardening Large Milling Cutters.

It is remarked that large, plain or former milling cutters, say over $3\frac{1}{2}$ in. in diameter and 4 in. long, to harden should be packed in a mixture of equal quantities of granulated charred leather and charcoal, taking care not to have any part of the mill within, say, 2 in. of the box at any point. Keep it in the furnace for four to four and a-half hours after the box is heated through to a low red ; remove the box from the furnace at the expiration of the time and quench the cutter in a bath of raw linseed oil, twirling it around rapidly in the oil so as to cause the oil to come in contact with the teeth. Allow the cutter to remain in the oil until cold. A formed mill with heavy teeth does not need to have the temper drawn. Mills with teeth cut in the ordinary manner should be run quite as long, and may be drawn for ordinary work to a light straw colour, or if drawn in a kettle gaging the heat by a thermometer to 425 or 430 deg. F. We have seen a large number of milling cutters and similar tools treated by this method and have never known one to be lost by cracking. In years of experience with this method we have had but a few pieces crack.

Mr. Woodworth's work is thoroughly practical, and its value is greatly enhanced by numerous illustrations and an excellent scheme of arrangement.

---

## COMING EVENTS—MARCH.

**3rd.** - Civil and Mechanical Engineers' Society : General Meeting, Caxton Hall at 8 p.m —Birmingham Engineering Society : Visit Bumsted and Chandler's Works, Hednesford

**5th.**—Birmingham Association of Mechanical Engineers : Monthly Meeting.

**7th.**—Society of Engineers : Meeting at the Royal United Service Institution. — Edinburgh University Chemical Society : General Meeting — Institution of Marine Engineers : Meeting at Stratford, E., at 8 p.m.—Society of Arts : Cantor Lecture.—Birmingham Association of Mechanical Engineers : Visit Messrs. Thompson's Boiler Works, Ettingshall.

**8th.**—Junior Engineering Society : Meeting at the G.W.R. Mechanics' Institute, Swindon.

**10th.**—Institution of Electrical Engineers : General Meeting.—Association of Birmingham Students of Civil Engineers meet.—Birmingham University Engineering Society meet.

**12th.**—The Manchester Association of Engineers : General Meeting.

**14th.**—Society of Arts : Cantor Lecture.—Edinburgh University Chemical Society : General Business Meeting. — Institution of Mechanical Engineers : Graduates Meeting at 7-30 p.m.

**15th.**—Junior Engineering Society : Meeting at the G.W.R. Institution, Swindon.

**16th.**—Institution of Civil Engineers : Annual Dinner at Lincoln's Inn Hall—Sir W. H. White presides.— Birmingham Local Section Institution of Electrical Engineers : Visit British Thomson-Houston Company, Rugby, followed by Meeting in the University.

**17th.**—Institution of Mining and Metallurgy : Meeting at 8 p.m.—Birmingham University Engineering Society, General Meeting,

**18th.**—Institution of Mechanical Engineers : Meeting at 8 p m.—City of London College Science Society : General Meeting.

**19th.**—North-East Coast Institution of Engineers and Shipbuilders : Meeting of Graduate Section at Sunderland. — Staffordshire Iron and Steel Institution : General Meeting.

**21st.**—Society of Arts : Cantor Lecture.

**23rd.**—Institution of Naval Architects : Annual Dinner at 7.15 p.m.

**24th.**—Institution of Electrical Engineers : General Meeting. — Birmingham Local Students' Institution of Civil Engineers : General Meeting. — Institution of Naval Architects : Annual Meeting,

**25th.**—North-East Coast Institution of Engineers and Shipbuilders : General Meeting.—Institution of Naval Architects : Annual Meeting.

**26th.**—The Manchester Association of Engineers : General Meeting.

**29th.**—Junior Engineering Society : Annual General Meeting at the G.W.R. Institution, Swindon.

**31st.**—Leeds Association of Engineers : General Meeting.

# AUTOMOBILE DEVELOPMENTS.

## By J. W.

It is announced that an automobile society has been formed in Calcutta, which will be known as the Automobile Association of Bengal.

It is interesting to note that the Berlin postal authorities are making use of motor vehicles for the conveyance of mails between the general post office and the suburbs of Schöneberg.

The side slip trials arranged by the Automobile Club are to be held about March 25th, and it is announced that a private track has been kindly lent for the purpose. The competition must necessarily involve a great deal of trouble to the organisers, and it will be interesting to see how the track is treated for this unique competition.

On March 19th to the 26th, Cordingley's Ninth International Automobile Exhibition will be held at the Royal Agricultural Hall, London. There will be ample opportunity to compare British machines with foreign productions, and an interesting feature will be a collection of war balloons and air-ships, which will be exhibited by the Aero Club.

### Motor Vehicles for Agriculture.

The practical application of motor vehicles to agriculture is well illustrated in a motor manufactured by the Ivel Agricultural Motors, Ltd. This is expressly designed for operating ploughs, mowing, reaping, chaff-cutting, churning, etc. It can also be used as a tractor or for stationary work, such as chaff-cutting, pulping roots, and grinding corn. The motive agent is petrol, and the machine, which is simple in construction, develops about 8 h.p.

O'Gorman's Motor Pocket Book* is a work which is to be cordially recommended not only to those who habitually use the automobile in one form or another, but also to engineers who desire to keep themselves well informed as to the progress made in the technicalities of the subject. Those who are newly taking up motoring will find it a veritable vade-mecum, for it is arranged on the dictionary plan, and it would be hard to name a subject useful to the motorist, from accumulators to wire gauges, that is not succinctly dealt with.

### The New York Show.

There was a large display of electric vehicles at the recent automobile show at Madison Square Garden, New York. Out of a total of 250, nearly 50 were of the electric class ; 175 were gasoline, and a dozen were steam motors, the others being also gasoline but of the motor bicycle type. An electrical contemporary remarks that there is nothing in such figures to make electrical engineers despair of the future, especially as a large proportion of the electrics were not of the fancy type but were strictly industrial and utilitarian.

The show, which was held under the auspices of the automobile Club of America, was not remarkable for any particular novelty, but it showed a marked advance over its predecessors. A great many of the electric automobiles were of the commercial type.

* Published by Archibald Constable and Co., Ltd. Leather, 7s. 6d. net

### The International Motor Boat Contest.

We are informed by Mr. Basil H. Joy, technical secretary to the Automobile Club, that three challenges were received from France for the British international cup for motor boats before the last day on which a challenge could be received. Two of these entries are for boats driven by petrol motors, one from MM. Clement and the other from Messrs. G. Pitre and Co., and the third is a Gardner-Serpollet steam launch, entered by MM. Legur and Gardner. No less than seven boats have been entered to defend the cup on behalf of Great Britain, two being from Mr. S. F. Edge, the present holder, three from Messrs. J. E. Hutton, Ltd., one from Messrs. Thornycrofts, and one from Lord Howard de Walden. This will necessitate an eliminating race being held to decide upon the three boats which are to represent England in the race itself. Further entries, it is hoped, will yet be received from France, which will also necessitate an eliminating test, and if, as is confidently expected, entries are received from Germany and the United States, the race will be the most representative and important international contest for motor boats that has ever taken place. The actual date of the race, which will be held probably in the Solent, is to be July 30th.

### The Crystal Palace Show.

The remarkable progress of the English automobile industry was well illustrated at the Crystal Palace Show, where a space of 130,000 square feet was eagerly taken up by motor manufacturers and traders ; in fact, it is impossible to avoid feeling that our home manufacturers have scored a triumph. The excellent show of motor agricultural machinery must have been particularly instructive to visitors from the country, and should lead to further important developments in this direction, while another very interesting feature was to be found in the steel motor boats. The *Times* calls attention to the fact that the British manufacture has practically doubled itself during a year by no means remarkable for commercial activity, and that a number of the great engineering firms of Great Britain have taken in serious earnest to the production of motor cars. Amongst them, without attempting an exhaustive list, may be mentioned Sir William Armstrong, Whitworth and Co., the Hozier Engineering Company, Messrs. Croxley, Messrs. Ransome—the last-named bring the motor to the aid of the gardener—and Messrs. Thornycroft. Side by side with these, the firms which live for the production of motors only—it would be invidious to name a few in this connection, and it is impossible to name them all—have redoubled their efforts , and the satisfactory result is that the proportion of British to foreign exhibits is far larger than heretofore. For ingenuity, variety, and engineering finish, the British exhibits certainly need not fear comparison with any that are foreign. In fact, there is no reason why a purchaser who would rather see his money go to his fellow-subjects than to foreigners should seek the aid of Continental or American manufacturers except that, rapidly as the British trade has increased, it has by no means kept pace with the British appetite for motor vehicles. All this is very encouraging for the motor industry, and we venture to hope that the "appetite will, nevertheless, be largely met at home. The exhibition included 270 stands and over 1,000 cars were shown, in addition to heavy vehicles and boats, and innumerable accessories.

# AMERICAN RÉSUMÉ.

## BY OUR NEW YORK CORRESPONDENT.

NEW YORK, February 19th, 1904.

### A Large Guillotine Shear.

The largest guillotine shear ever built by the Hilles and Jones Company was recently installed in the warehouse of Joseph T. Ryerson and Son, Chicago. The upper blade is 13 ft. 2 in. long and will shear a plate 1½ in. thick and 12 ft. long, working between the housings. The overhang or gap is 36 in., which feature permits a strip 36 in. wide or less to be sheared from the side of a plate of any length, the plate, of course, being moved along parallel with the blade after each stroke. The shear complete weighs 220,000 pounds. It is driven by a 50 h.p. Westinghouse three-phase motor, direct connected by gearing and a friction clutch. The frame stands 21 ft. above the floor and the total width is 22 ft.

### The Chicago Fire and After.

The terrible fire at Chicago may not after all be without its advantages. Architects and builders, who make it a business to remodel buildings, are extremely busy in Chicago and the West now, as a result of the rigid enforcement of rules and ordinances for fire protection. Not only theatres and halls are undergoing reconstruction, but many churches and other places of assembly will have to be completely remodelled. Every manufacturer of fire escapes and steel and iron stairways is crowded with rush work. Thousands of workmen are finding employment, and it is believed that in Chicago alone $1,000,000 will be expended for rendering buildings safe that have not been sufficiently so.

### American Pig Iron.

James M. Swank, general manager of the American Iron and Steel Association, has published his statistics of the production of pig iron in the United States for 1903, the report being based on complete returns from every furnace in the country.

The total production of pig iron in 1903 was 18,009,252 gross tons, against 17,821,307 tons in 1902, 15,878,354 tons in 1901, 13,789,242 tons in 1900, 13,620,703 tons in 1899, and 11,773,934 tons in 1898.

### 1904 Electrical Congress.

Over eight hundred have already accepted membership of the 1904 International Congress, which European engineers are now being invited to join. As some of the best electricians in Europe and America have promised to contribute papers to the sections in September, it is anticipated that the transactions will have an exceptional electro-technical value.

### Producer Gas from Wood.

A most novel feature of the power plant of the Montezuma Copper Company at Nacozari, Sonora, Mexico, as described by Mr. John Langton to the American Institute of Mining Engineers, is the use of producer gas made from wood alone. No guiding experience was

found for this process ; but, with the desire to utilise as far as possible the limited local wood-supply, the gas-producer plant was selected with the object, among other things, of determining the advisability of using, if not wood alone, at least a considerable admixture of wood with bituminous coal. The power-plant was started in operation July 31st, 1900, with coal ; but for some months there was no convenient opportunity to experiment with wood, and it was not until February 4th, 1901, that the first trial of wood alone was made.

The most obvious difficulty to be feared arose from the large proportion of condensible distillates yielded by wood, and the danger that some portion of these might be imperfectly fixed in passing through the producer. The trouble from tar deposited in the gas apparatus and pipes would be serious, and even a small quantity of tar in the gas itself is a fertile source of trouble at the engine valves. Unless a permanent gas could be made from wood, this fuel would be unavailable.

The first care, therefore, was to insure that there should be a bed of charcoal on the grate sufficient to form an adequate fixing zone. To obtain this the producers were filled about 5 ft. deep with cordwood sawn in blocks about 6 in. long, and the contents blown with a slow fire for four or five hours before the gas was turned into the holder. The gas, as it proved, was turned into the holder too soon. At first, it contained some tar and it was not until after about three hours' operation that the charcoal accumulated in sufficient quantity, so that the producers delivered fixed permanent gases to the holder. This was evident on exposing white paper to the gas.

This preliminary trial lasted more than twelve hours, until the supply of sawn wood was exhausted. As was expected, upon drawing the fires a considerable quantity of charcoal was obtained, which was available for use as a bed in starting the succeeding fires. But before the next trial was made it was decided that for this purpose coke would be better than charcoal, and the practice has been continued ever since. The experience at Nacozari, however, does not determine whether it would be safe to dispense with the use of coke.

The second trial, beginning March 14th, was carried on for 66 hours with perfect success ; and after a third trial of 36 hours' duration, made a few days later, had proved that there was no difficulty in repeating the previous excellent results, measures were taken to gather a supply of wood sufficient to operate the plant with wood-gas exclusively.

The wood used is principally scrub white oak, with about 10 per cent. ash.

### The American Civil Engineers.

At the annual business meeting of the American Society of Civil Engineers at New York, the most important matter taken up for discussion was in regard to accepting the proposition of a joint building for the various national engineering societies as a gift from Mr. Andrew Carnegie. It was decided to instruct the Board of Direction to issue a letter ballot to the membership, to be canvassed at the meeting of the society on March 2nd.

It was shown that 212 members had been added to the roll during the year, bringing the total to 2,924, of whom 718 are resident members and 2,206 non-resident.

## Mining Yellow Ochre.

A paper recently presented by Mr. Thomas Leonard Watson to the American Institute of Mining Engineers, dealt with the " Yellow Ochre-Deposits of the Cartersville district, Bartow County, Georgia."

The principal use made of the Cartersville ochre at present is in the manufacture of linoleum and oil-cloths. For this consumption the principal markets are in England and Scotland, to which the bulk of the Cartersville product is exported. Some of it is used in the United States for a similar purpose. It is also used to a limited extent in the manufacture of paints. By calcining, the ochre is converted into a desirable dark-red pigment.

The ochre-deposits to be mined in the Cartersville district form extremely irregular branching veins, which intersect the rock in almost every direction. The ore-bodies may occur enclosed in the hard and fresh quartzite, or they may be entirely enclosed in the residual decay derived from the quartzite. The bodies of pure ochre are usually soft and clay-like in character, and the ore is easily mined with the pick and shovel. They are generally exposed along the slopes and summits of the quartzite hills and ridges.

On those properties in the district where systematic mining has been done, the method employed consists of tunnels driven into the ridge, from which drifts are worked at suitable points. In this way a number of levels have been worked, one above the other, on several of the properties. Occurrence of the ochre in the fresh rock, as on the Georgia Peruvian Ochre Company's property at the wooden-bridge across the Etowah river, at times necessitates blasting.

Timbering is necessary in the tunnelling, as caving is apt to occur. The underground-mining is also extensive enough to necessitate tramways and lights. The tram-cars are drawn in and out of the mines either by mules or by means of steam and cable. Both are in use in the Cartersville mines.

## Steel Cars for New York Subway.

According to the *Iron Trade Review*, the American Car and Foundry Co. has contracted with the Interborough Co., of New York, which will operate the underground rapid transit and leases the Manhattan Elevated, to furnish the underground railway with what are said to be the first steel non-combustible passenger cars ever built

or designed. The contract involves about 1,500,000 dol., and calls for 200 cars. The design of the new car was worked out by George Gibbs, consulting engineer of the Interborough Co. These cars will be built at the Berwick plant of the American Car and Foundry Co. The cars are to be delivered next summer. It is pointed out that owing to the frightful accident in the Paris tunnel last summer the Interborough Co. has desired to spare no expense in order to avoid accidents to passengers. When the subway starts in active operations there will be 500 cars put on. If the all-steel cars prove successful they will gradually replace the others.

## A Large Smelter Chimney.

From the standpoint of volume of gas discharged the new chimney at the Washoe Smelter, at Anacond, Mont., will be the largest in the world. Others are taller, but their diameter, and consequently their capacities, are smaller. When completed the base of this stack will be 45 ft. square, while the top will be 36 ft. in outside diameter, 300 ft. from the ground, and 6,182 ft. above the sea level. A flume 2,667 ft. long will convey the gases from the roaster to the base of the stack, and be the cause of precipitating much of the poisonous matter usually thrown into the air. Arsenic is one of these impurities and it exists in such large proportions in the smoke that plans for a possible refining of this as a by-product are under consideration. The chimney will be constructed of red brick and concrete, and is expected by its great elevation to carry the noxious fumes from the smelter to the upper air currents, where they may be thoroughly and harmlessly dissipated.

## Electric Cars in Korea.

On January 24 the United States State Department at Washington received official information of an attack by a mob of native Koreans on an electric car because it had killed a Korean. The news came in the following cablegram received from Minister Allen at Seoul : " This morning on the electric railway, which is the property of American citizens, a Korean was accidentally and unavoidably killed. Thereupon a mob of natives attacked and partially destroyed the car. The operators of the car would have been injured had it not been for the presence of mind and action of our guard, and a serious riot would have occurred." Although there have been previous reports of disturbances in Korea, says the *Electrical World*, this is the first mob attack made on the property of Americans. The railroad is owned and operated by Americans, H. R. Bostwick, of San Francisco, and H. Collbran being its principal officers. It runs through the heart of Seoul. The United States Legation guard now consists of 100 marines. The reinforcement of this guard has been urged, and could be effected in a week's time by details of marines from the Philippines.

# SOUTH AFRICAN RÉSUMÉ.

## BY OUR JOHANNESBURG CORRESPONDENT.

JOHANNESBURG, February 12th, 1904.

## Navvies in the Transvaal.

A large experiment in the employment of white unskilled labour has been terminated in an unsatisfactory manner during the last month by the dismissal of 963 navvies who had been employed for the preceding ten months on the construction of 20 miles of earthworks on the Springs-Ermelo line of railways.

The men, after a fairly good start, reduced their output until Mr. Wall, the Chief Constructor, visited the works, and, by dint of persuasion and the institution of a system of bonuses, induced them to renewed efforts which resulted in an average of 6¼ cubic yards per man per day for a time. But afterwards the amount of work gradually fell oft again until it actually fell to a little over 1 yard per man. This being reported to the authorities, the latter decided to stop the work and send the men back to England.

This decision cannot be wondered at, for, according to the writer's own experience of railway construction in this country, the average work done by Kaffirs in ordinary pick and shovel ground, on monthly wages, without any special inducements, ranges from 3¼ to 4 yards per day per native. Mr. Wall states that the extra expenditure on the 20 miles due to the substitution of white labour for black amounts to £75,000. This probably includes the cost of importation and other items which would not recur frequently if the men had worked well and remained in the country, but the other figures given show that white labour does not, and cannot, compete against black in South Africa as regards cost, unless the former receives what is a starvation wage, a state of things which nobody desires to see established.

It has been stated that " the navvies did not desire to acquire efficiency or to make their work valuable,'' and, again, that " they would not work,'' but such bald, unsympathetic statements do not take sufficient account of the climatic conditions, and the effect on a white man of severe and continuous manual labour out in the open for prolonged periods. So that, although this failure tends to further prove that white unskilled labour cannot be profitably employed on a large scale, it does not necessarily show that the fault lay with the particular men in question.

## The Passing of the Rand Central Ore Reduction Company.

Had this company been merely a dividend-earning concern its decision to go into voluntary liquidation would call for no comment in these columns. But it has been so closely associated with the expansion of metallurgical practice on the Rand that its career has been of considerable technical interest.

Formed in 1892 for the purchase and treatment of concentrates and large heaps of tailings, it had in June, 1894, six separate cyanide plants at work on sand treatment and one chlorination plant for concentrates.

As, however, the advantages and methods of cyanide treatment became more widely appreciated, the mining companies preferred to build their own plants instead of selling their tailings. The Rand Central, therefore, diverted its energies necessarily into new channels, and besides constructing many of the new plants for the mines, also experimented upon the treatment of slimes. About the same time it introduced the Siemen's electrolytic method of precipitation, and established a refinery for the treatment of the lead bullion produced by this process.

In 1897 a blast furnace was put in operation for the production of pig lead and silver from the galena ores mined near Pretoria. This was followed by the erection of mills and presses for the manufacture of lead foil, sheet lead and pipes.

The principal factors which have led to the proposed liquidation have been officially given as follows : The unfavourable results in many instances obtained from slimes purchased by contract, the unexpected difficulties encountered in the treatment of old accumulated slimes. The enforced idleness of the plants during the war. The decrease in quantity and value of the by-products purchased from the mines which has resulted from improved methods and more careful work on the part of the mining companies. The abolition of the duty of £25 per ton on imported lead, coupled with the present scarcity of lead ore and the high cost of its transport.

## Precautions against Gassing.

A member of the Chemical Metallurgical and Mining Society of South Africa sets forth in the journal of that society the recommendation that every mine be provided with a cylinder of oxygen and fan piece for use in cases of "gassing."

The elimination of the CO from the blood and tissues begins to take place directly the person is removed from the poisonous gas into the ordinary air. As decomposition of the carboxyhaemoglobin is the result of a mass action depending upon the relative quantities of oxygen and carbon monoxide brought into intimate contact with the haemoglobin, and as the affinity of carbon monoxide for haemoglobin is very much greater than that of oxygen for haemoglobin, the greater the amount of oxygen which can be brought into contact with the blood in a given time the more rapidly will the carbon monoxide be turned out of the blood.

When a person who has been even partially gassed has to recover by breathing ordinary air, each breath he takes contains four-fifths of the diluent nitrogen, so that only one-fifth of the volume of air he inhales enters into the reaction. It is obvious, therefore, why absolute recovery is so prolonged. If, however, pure oxygen be inhaled in lieu of air, the volume of oxygen actually reaching the lungs at each inspiration being so much greater than when ordinary air is breathed, the carbon monoxide will be driven out of the haemoglobin at a much greater rate.

# GERMAN RÉSUMÉ.

BY

Dr. ALFRED GRADENWITZ.

BERLIN, February 22nd, 1904.

## A Novel Electric Traction System.

An electric railway traction system of quite a novel kind is being developed at the present moment by a Swiss " Studiengesellschaft," appointed for the purpose of finding out an electric railway system suitable for that country, which on account of her dependency on the foreign coal market should endeavour to utilise her wealth of hydraulic power. Speeds, on the other hand, are limited there because of the steep gradients, sharp curves, and numerous stoppages. In the system in question, as described in a recent issue of the " Elektrotechnischer Anzeiger," steam locomotives heated by electricity are adopted. Electric heating, as is well known, will work with the highest possible efficiency, so that the total efficiency mainly depends on the output of the mechanical part of the locomotive, viz., the steam engine proper. Any coal steam-locomotive could readily be converted into an " electrothermic " locomotive by simply replacing the fire-box and boiler tubes by a number of parallel electric heating walls running throughout the boiler, and consisting of two copper or iron sheets. It is suggested using in this connection the well known Prometheus heating elements. The consumption of current would depend on the consumption of steam ; let the boiler be designed for accommodating 4,000 litres of water, which are to be brought within three hours from 10 to 190° C., corresponding with a steam pressure of 50 kg. per sq. cm. ; in the case of an efficiency only as high as 90 per cent., the following data would be obtained : 4,000 litres of water would require, in order to be brought to the above temperature, 4,000 by 180=720,000 kg. cal. ; 1 kg. cal.=1,275 eff. wt. hrs. ; therefore, 720,000 kg. cal. =about 900 eff. kw. hrs., or distributing this amount over three hours=about 300 kilowatts.

A consumption of steam of 1,000 kilogrammes per hour would accordingly require a supply of current of about 225 kilowatts.

As regards the advantages inherent in the electrothermic system, the resistance of the steam accumulator against current shocks should be mentioned. There is the further advantage of both direct and alternating currents being practicable in this connection, and any desired combination being possible. The mean efficiency of electrothermic locomotives, being about the same as that of an electromotive machine of the same size, would be about 60 per cent. to 70 per cent., whereas the total efficiency of the railway system, on account of the more advantageous utilisation of the load, would be higher for the former. Furthermore, the adoption of electrothermic service may take place gradually, being much easier than that of electromotive service, on account of the lower cost of the conversion and the easiness with which the personnel may be trained for the new service. A possible conversion of electrothermic into electromotive railway service would finally be readily made, should the electromotive service in future be so improved as to become superior to the electrothermic system.

## Rational Modifications in the Design of Marine Engines.

At a recent meeting of the Hamburg section of the *Verein Deutscher Ingenieure*, Mr. Freytag suggested some modifications in the design of marine engines as derived from the progress obtained in designing stationary steam plants. The main requirements would be high figures for the steam tension and the number of revolutions in connection with high linear speeds and strongly superheated steam. These requirements it would not be possible to fulfil without using such governing devices as would exclude any noxious effects of the enormous pressure of superheated steam. Using the slide in this connection would be quite out of the question, whereas the valve, on account of the even distribution of its mass, preventing any noxious effects due to thermic expansion would be quite suitable, the consumption of power being rather low on account of the small resistances. The valve would, moreover, be specially suitable for separating the inlet and outlet, this ensuring a high independency in distributing the steam and choosing the driving device of the distributor. As, moreover, the weight of the valve is quite immaterial as compared with the weight of the slide, there would be no mass and weight effects due to the valve ; nor would water shocks be as likely to occur in the case of valve machines as in the case of slide engines. A guided valve distributor of a simple design with safety valve effect and without stuffing boxes, it is pointed out, would best meet modern requirements. Lentz distributors, as well as any analogous designs, would be quite suitable, the coal consumption of machines provided with similar distributors being much less than the one of the usual modern ship engines.

## Comparative Tests of Incandescent Gas and Electric Arc Lamps.

The Schuckert and Co. Electric Company, Nürnberg-some time ago, caused comparative tests of incandescent gas and electric arc lights to be made, the results of which have just come to hand. These experiments, as conducted by Dr. Lehmann Richter, gave the following results : The surface luminous intensity at the level of the eye proved fully satisfactory for both light sources, arc light affording also a uniform distribution of light. In the case of electric arc light, no noxious alteration of the air was noted ; the temperature would not rise to any appreciable extent, nor would the percentage of carbonic acid be augmented. With incandescent gas light on the other hand, the temperature at the level of the eye was found to rise by about 68° C. in the course of three hours, while the percentage of carbonic acid was found to increase up to a value more than five times the initial figure. As regards the cost of operation of both classes of light, this proved somewhat smaller in the beginning with Auer light, whereas after a short time the figure corresponding to arc light was reached even without taking the lighting flame into account. When accounting for the lighting flame, on the other hand, the cost of Auer light would be higher than that of electric arc light.

## Comparative Experiments with Saturated and Moderately Superheated Steam in Locomotives.

In the course of some experiments made on behalf of the Breslau Royal Railway Department, and described in a paper by Mr. Strahl (see *Zeitschrift des Vereins Deutscher Ingenieure*, No. 1 2), two high-speed train locomotives were fitted with Pielock super-heaters, and compared with two similar locomotives without superheater as to their consumption of water and coal. The main results arrived at may be summarised as follows :—

1. The temperature of the steam on issuing from the superheater being 260°, the saving as to water vaporised was about 16 per cent., and as to the consumption of coal, 12 per cent. ; whereas the steam-saving proved equal to about 10 per cent. for a mean steam temperature of 230° in the dome.

2. The consumption of steam in the different locomotives compared, proved the same for equal outputs.

3. The weights of water vaporised are inversely as the specific volumes of the different kinds of steam being directly proportional to the specific weight. The saving in steam obtained corresponded with the increase in the specific volume of the steam due to superheating.

4. The saving in steam, being dependent only on the superheat, must be the same both in compound and twin locomotives for equal superheat, the comparison being relative to quite similar locomotives.

5. Slide valves could be used up to the highest temperatures attained (272° C.) provided sufficient oil was supplied to the sliding surfaces by means of lubricating presses.

6. In order to fully utilise the advantage inherent in superheating for a higher efficiency of the locomotive, the cylinders should be increased proportionately to the higher consumption of heat (coal) of the locomotives (in the case of equal outputs) without superheaters, as against locomotives with superheaters.

### High-Speed Steam Railway Service.

As an indirect consequence of the Marienfelde-Zossen high-speed electrical railway trials, experiments are being made on a number of German railway lines with a view to investigating the working conditions of a steam railway service with increased speeds. On the Cassel-Hanover line, for instance, the trains tested are made up of gigantic high-speed locomotives and solidly connected six-axle cars, warranting a mean speed as high as 130 kilometres (81 miles) per hour. This speed would enable the journey between Berlin and Hamburg to be completed in about two hours, and it is safe to state that one such train in either direction would be quite sufficient for the present traffic. In the case of these experiments giving satisfactory results, it is thought probable that next summer some specially suitable lines will be arranged for a similar increased speed service, the more so as the Berlin-Zossen trials have shown existing permanent ways (provided they be fitted with heavy rails) to be fully suitable for a similar service. Even in the event of the introduction of electric high-speed railways being postponed for economical reasons, a material improvement in the German high-speed railway service may be anticipated, as far as lines with specially dense traffic are concerned.

### Optical Telegraphy.

Our readers will doubtless remember the beautiful experiments in wireless telephony which were made by Herr E. Ruhmer on the Wannsee Lake, near Berlin, last year, and continued with increasing success in the course of last summer. Now the inventor has applied his process to optical *telegraphy*, and the

Siemens-Schuckert Werke are just now bringing out this novel wireless telegraphy apparatus.

In optical telegraphy the rays issuing from a projector are, as a rule, intercepted at given intervals, so as to form luminous flashes, succeeding one another more or less rapidly. In the Ruhmer telegraph system, on the contrary, the so-called *speaking arcs* are utilised by superposing on the continuous current circuit of the lamp placed at the sending station in the focus of a projector, a continuous current frequently broken by means of a mechanical interrupter, the opening and closing being ensured by a Morse key, in accordance with ordinary Morse signals. At each closing of the telegraph key the superposed and frequently interrupted continuous current will modify the luminous intensity emanating from the electric arc, giving rise to luminous oscillations which are projected towards the receiving station. If all the conditions be so arranged that the luminous intensity of the lamp is maintained constant, this process will ensure not only a more rapid handing of telegrams, but will permit at the same time of keeping the latter strictly secret, as the human eye, incapable of discerning any more than ten luminous alternations per second, will get the impression of a continuous beam on account of the rapidity with which the luminous oscillations of the transmitting station will succeed each other.

The receiving station is arranged in a way analogous to those of optical telephony, comprising two telephones and one parabolic reflector in the focus of which the selenium cell is placed. The luminous oscillations of the transmitting station are perceived in the telephone of the receiving station by means of the selenium cell as humming intermittent sounds, constituting *acoustical* and directly perceived *Morse signals*. The pitch of this sound will depend on the frequency of the interrupter. Whereas in transmitting language, uncertainties are possible on account of the different acoustical intensities of the different vowels, the same sounds have to be heard here for more or less prolonged intervals. It has therefore been possible to ensure perfectly clear transmissions of signals in atmospheric conditions which would have rendered difficult the transmission of language. The beginning of a communication is indicated by a bell, operated by the selenium cell without the agency of any wire connecting it with the transmitting station.

The satisfactory results of the experiments so far made go to show that this system of optical telegraphy, like the analogous system of optical telephony, will be used to special advantage in the case of transmissions over brief distances. It will, therefore, be especially suitable for military and naval purposes.

### The Berlin Telephone System.

According to the latest returns, the number of subscribers to the Berlin telephone system has reached the enormous figure of 73,000. This would mean an increase by about 10,000 in the course of last year. Of the above total, about 11,000 would be relative to the suburbs, Berlin itself being represented by about 62,000, *i.e.* 6,000 subscribers more than last year. The relative increase in the number of subscribers in the Berlin suburbs being about 4,000, is thus much higher. The largest Berlin telephone exchange is the main telephone office, comprising 12,340 connections.

# THE MINING WORLD.

### By A. L.

THE world's copper output in 1903 was 589,361 long tons, compared with 551,316 tons in 1902 and 515,992 tons in the previous year.

Colonel K. M. Foss, in a recent interview with a *Times'* correspondent, said he had been working northward of Mergui, in Lower Burma, to the Siamese frontier with engineers, and had discovered the existence of deposits of tin ore equal to those in the Straits Settlements and likely to add largely to the world's output. Excellent coal was also found in the vicinity.

The American Society of Civil Engineers has appointed a Committee to represent the Society at the Universal Exposition to be held at St. Louis in 1904. The Headquarters will be located in the Liberal Arts Building, and the Members of the North of England Institute of Mining and Mechanical Engineers have been cordially invited to avail themselves of the conveniences to be provided during the Exposition. The Society will exhibit a collection of plans, photographs, models, and descriptions of American engineering works, which they have no doubt will prove of great interest, and in addition they intend to make their headquarters a centre of information as to all other exhibits which may be of engineering interest. A register will also be provided, and it is hoped that the headquarters may serve as a rallying point and rendezvous for all visiting engineers.

### Electricity in Mines.

Since the last issue of PAGE'S MAGAZINE went to press the report of the Departmental Committee on the use of electricity in mines has been issued as a blue book. As general principles which should govern the employment of electricity in mines they suggest the following :—

(1) The electric plant should always be treated as a source of potential danger ; (2) the plant, in the first instance, should be of thoroughly good quality, and so designed as to ensure immunity from danger by shock or fire, and periodical tests should be made to see that this state of efficiency is being maintained ; (3) all electrical apparatus should be under the charge of competent persons ; (4) all electrical apparatus which may be used when there is a possibility of danger arising from the presence of gas should be so enclosed as to prevent such gas being fired by sparking of the apparatus ; when any machine is working, every precaution should be taken to detect the existence of danger, and on the presence of gas being noticed, such machines should be immediately stopped.

General principles are formulated with regard to the application of electricity, in reference to generating stations, cables, switches, fuses, etc., stationary motors, portable motors for coal cutters, drills, etc., electric locomotives, electric lighting, shot-firing, signalling, and electric relighting of safety lamps. It is pointed out that there may be parts of mines in which so much fire-damp is emitted that the introduction of electricity might be improper, notwithstanding the fact that by law these places may be worked with safety lamps. The committee, therefore, emphatically impress on both colliery owners and managers that the ultimate responsibility of installing or using electricity in such places should and must rest with them. A very valuable set of suggested rules is appended.

With regard to the much discussed subject of pressure, the committee considers that a reasonable limit would be medium pressure not in any case to exceed 650 volts. The report is signed by Mr. H. H. S. Cunynghame, C.B. (Chairman), Mr. Charles Fenwick, M.P., Mr. W. W. Hood, Mr. W. H. Patchell, Mr. W. N. Atkinson (H.M. Inspector of Mines), Mr. A. H. Stokes (H.M. Inspector of Mines), and Captain A. Desborough (secretary). It should be of considerable service in dispelling some of the exaggerated notions which have been felt among mine-owners and managers as to the dangers of electricity in mines.

### Electric Light for Deep Levels.

A strong plea for the use of electric light in the mines of South Africa was made by Mr. T. L. Carter, a speaker who participated in the discussion which followed a paper on the ventilation of deep levels, read before the Chemical, Metallurgical and Mining Society of South Africa. It was pointed out that the bearing of mine illumination on the ventilation question is an important point to consider. Other things being equal, the illuminant which causes the least vitiation of the atmosphere should be adopted. The old paraffin flare lamp is one of the favourite methods of lighting up the stations. Anyone who has ever worked in a drive with a paraffin flare lamp a few hundred feet from a cross-cut, will have found that in less than a quarter of an hour either he or the lamp had to go out. On some mines acetylene lamps are being used instead of a flare lamp. One of the greatest advantages of electric lights in a mine is that they do not contaminate the mine atmosphere. When possible, therefore, electric light should be installed.

## Mining Developments in Korea.

At the present juncture it is interesting to turn to the paper on mining in Korea which was contributed a year ago by Mr. L. J. Speak at a meeting of the Institute of Mining and Metallurgy. Attention was called to it at the time in the columns of PAGE'S MAGAZINE. The author pointed out that Korea was not then open to foreigners for mining, with the exception that one subject of each of the Great Powers might secure one concession.

The principal terms on which these concessions were granted were that mining supplies might be imported duty free, and that the King should receive 25 per cent. of the profits.

Water is plentiful, except for a short period during the height of the winter. Lumber, mining timbers, and cordwood, though not too plentiful, are cheap owing to the cheap labour, but steps are now being taken to develop a water-power scheme in order to preserve the timber. Labour is generally plentiful, but considerable difficulty is met with in obtaining suitable white foremen and overseers, who are mostly obtained from the Western States under contract, and, as in similar cases all the world over where personal selection is not possible, are not always satisfactory. Japanese are largely employed as carpenters, blacksmiths, and engineers, and many of them are excellent workmen ; their wages are mostly three shillings per day, but a few get more. Chinese are largely employed as surface coolies in the mills and cyanide works, and to a limited extent underground. They are preferable to Koreans for such employment, as they work more regularly and require less supervision. They are also indirectly useful in preventing labour troubles and checking thieving, as they do not mix with the Koreans. The ordinary wage of a Chinaman is 10½d. per day. Koreans are employed for the rest of the work ; their carpenters are expert adze-men, and as miners and tool-sharpeners become very efficient. At a recent drilling contest, the winning double-handed team, using ⅞-in. steel drills, sharpened in the ordinary way, finished 22 in. in a granite boulder in ten minutes.

## Korean Coolie Labour.

The pay of an ordinary Korean coolie is 7½d. per day, and of a miner or carpenter, 1s. 3d. a day. No food or lodging is provided for any of the Oriental workmen. Koreans run most of the hoisting engines and no serious accidents have occurred. After allowance is made for the difficulties of language, it must be said that these Japanese, Manchurians, and Koreans are as intelligent and as capable of receiving instruction as a European would be who had been brought up without knowledge of our methods. Their religious and moral ideas are somewhat crooked, but they are amenable to common-sense. A Korean is not so conservative as a Chinese.

The main principle on which this labour is managed is to have all natives work under the direct supervision of white men without any intermediate native foremen. With proper organisation the number of labourers a white man can look after is mainly determined by the extent of ground they are spread over.

## The Mining of Corundum.

In the course of his fourth Cantor lecture on the Mining of Non-Metallic Minerals, Mr. Bennett H. Brough confined his attention to precious stones, one of the most interesting sections dealing with Corundum gems. The methods of mining for the ruby in Burma are suited to the three modes of its occurrence in the limestone, in hill detrital material, and in the alluvial deposits in the valleys. In the quarries blasting is unsuitable as it injures the gem stones. The dirt is raised by endless ropes from quarries 50 ft. deep. Stones of greater weight than four carats are of such exceptional occurrence that they command fancy prices. The largest known were brought from Burma in 1875 and weighed 37 and 47 carats respectively. They are said to have been sold for £10,000 and £20,000.

Emery, the common form of corundum, is found in large quantities in the Island of Naxos, and the mines are the property of the Greek Government. The Naxos emery is superior to that of Asia Minor owing to its great density, and the greater fineness and hardness of its grain. It fetches 112 to 115 francs a ton. The world's consumption of emery is 25,000 tons annually, of which Asia Minor supplies some 18,000 tons, valued at £53,000 ; Canada 388 tons, valued at £10,914 ; and Naxos 6,328 tons, valued at £26,830.

These lectures, by the way, have been bound up together, and are published by Wm. Trounce, of 10, Gough Square, Fleet Street, E.C.

## Mining in China.

In view of the enormous mineral resources of China and the equally enormous supply of labour, it is impossible to avoid the conclusion that this much-disputed region is destined to play a most important part in future mining developments. A correspondent of the "Engineering and Mining Journal" lately pointed out that were it not for Government control the antimony mines which are now being worked in China would swamp the market. As it is, prices of refined antimony have been forced down over one-third, and Chinese antimony ores are offered of good quality and at low prices. The mines are worked by hand labour, with crude tools, and no machinery is used. The ores, as brought to the surface, are sorted by women and children, and are broken, where necessary, into small pieces, hand-culled and dressed. In this way the ore for export and for sale is practically pure. The rejected ores are smelted into crude antimony by a primitive method with great waste. The method is by volatilisation, and the slags carry 25 to 30 per cent. metal. Under these conditions, with over 1,000 miles inland transportation, Chinese are able to ship quantities of both ore and regulus at a satisfactory profit. These conditions will probably apply to the working by hand labour of many other minerals, especially if, under practical supervision of foreigners, modern tools and methods of mining are used.

# OPENINGS FOR TRADE ABROAD.

## Straits Settlements.

A correspondent writes as follows :—
" I shall be much obliged if you will let me know
if I can procure a machine for injecting or pumping a
deadly gas into the earth, or white ants' mounds, to
destroy this pest. Termites (white ants) generally
live in mounds about 1 ft. or 2 ft. below the surface
of the earth. A hole might be made in the spot with an
iron bar of, say, 1½ in. diameter, then a hollow tube per-
forated at the end put in and the gas pumped into the
earth. I shall be much obliged for the above information and the cost of the machine and the gas."

## Spain.

Tenders, which should be delivered before March 14th
at the Secretariat of the port works of Castellon, are
in demand for the supply of a locomotive and tender,
suitable for a gauge of 1·30 metres, with three pairs
of wheels coupled, and of a maximum running weight
of 22,500 kilogs. A provisional deposit of 2,000
pesetas, or about £63, is required to qualify any tender.
An opening occurs in the Municipality of Valencia de
Alcantara, for the work of supplying and laying iron
tubing to replace the existing aqueduct which brings
the water supply into the said towns. Upon adjudica-
tion of the contract, the contractor has to deposit the
sum of 5,000 pesetas, or about £150.
Tenders are in demand which will be opened on
the 6th April next in the Ministry of Public Works,
Madrid, for the concession to construct and work for
sixty years a steam tramway in Valencia, from a station
in that city to the port of Valencia.
Concessions have been granted for a metre-gauge
railway from Barcelona to Junquera *via* Palamos,
with the following possible branch lines : From San
Adrian de Besos to connect with the narrow gauge
line from Igualda to Martorell and with the tramway
from Manresa to Berga ; from Lloret de Mar, to connect
with the narrow gauge line from Olot to Gerona ; and
from Vilamalla to Rosas and Olot. Also a narrow gauge
line from Berja to Ugijar, passing through Alcolea and
Cherin, with a branch to Canjayar.

## Russia.

There is a steadily growing demand for bicycles
and motor cars. In the latter case, the trade has
developed surprisingly in the space of a few months.
Manufacturers must bear in mind that, owing to the
roads, machines must be of the most solid type, and in
order to compete with cheap German and Belgian
makers, a depôt in Odessa, where detached parts could
be put together, would be of the greatest service, as
the duty on manufactured machines is, proportionately, much heavier than that on parts.

## Canada.

The present year promises an active development of
telephone competition in Canada. Several of the larger
cities of Ontario will have before them application
for franchises to compete with the Bell Telephone
Company. It is reported that a company have
recently obtained a Dominion charter, and propose
to build exchanges and construct long-distance lines
throughout Canada in competition with the Bell Company.

## Australian Commonwealth.

We learn from the Canadian Commercial Agent at
Sydney that in the Australian markets there should
be a good opening for heating apparatus, particularly
in churches, schools, and shops, beyond the coast
line, and there is no doubt that a manufacturer of
heating apparatus would find a good trade. It would
be necessary to send out an expert in hot air and hot
water, and open a warehouse ; though this need not
be an expensive one, it would require considerable
capital to do the business profitably.
He also reports the extraordinary success of artificial
irrigation on comparatively small areas of land, and
predicts for it a great future. The prospects for
selling pumping machinery, motors, and other power
generators—as, for instance, windmills and steam,
oil, and hot-air machinery—are very good. It would
be advisable to sell such machinery through agents
who understand how to put it up. Some firms that
sell that class of machinery for certain manufacturers
could be induced to deal in new ones, if good and cheap.
It would be necessary to send a competent expert and
put up some sample machines, in order to prove their
efficiency.

## Natal.

The Government of Natal is prepared to receive
tenders for the supply, erection, and completion of
plant capable of dealing with 4,000 tons of coal in ten
hours, consisting of one fixed and two movable appliances to discharge into ships' hatches direct from
trucks, or from storage bins of 10,000 tons capacity.
Also plant capable of emptying trucks and delivering
800 tons of coal per hour from four mechanical appliances, each designed to deliver the coal into ships'
coaling ports, or into hatches on deck for a single
line of ships up to 50 ft. beam, or, alternatively, up to
a distance of 70 ft. from the wharf face, so as to serve a
double line of ships.

## Mexico.

A contract has been arranged between the State
and Messrs. Pedro Ruiz y Manuel L. de Guevara for
the construction and working for ninety-nine years
of a railway in the State of Vera Cruz, from a point of
the Estero de Santecomapan, to Caleria, Canton de los
Tuxtlas, to be called the Tuxtlas and Golfo S. A.
Railway, with the right to prolong the line to San
Andres and Santiago Tuxtla. The gauge is to be 914
millimetres. The free import of materials and goods
referred to in Article 74 of the Railway Law will be
allowed for three years.

## Brazil.

The Budget for the present year authorises the
Executive to carry out the construction of the Port of
Belem, adopting the plans suitable for the portions
which have to be constructed between the bridge
of the naval arsenal and the Port of Pinheiro.
The Government has been authorised to construct a
railway from Timbó, in the State of Bahia, to the city
of Propria, in the State of Sergipé. This railway is to
join the cities of Aracajú and Simao Dias directly or
by the means of branches.
A decree has been published approving the estimate
of 2,722,107 milreis, or about £136,105, for the construction of the works of the first 60 kilometres which
constitute the first section of the extension of the
Central Railway of Brazil, between Curvello and the
San Francisco River.

# WHAT OUR TECHNICAL COLLEGES ARE DOING.

## By A TECHNICAL STUDENT.

### TECHNICAL EDUCATION AT DERBY.

FROM the Principal of the Municipal Technical College of the Borough of Derby, Professor F. W. Shurlock, B.A., B.Sc., we have received an imposing list of successes obtained during the session 1902–3, together with the calendar of the college. The half-tone illustrations give an excellent idea of the practical work carried on at the college, and some of them make one envy the students at Derby, who are evidently so well looked after.

### AT LIVERPOOL UNIVERSITY.

The work carried on by the Faculty of Engineering of the University of Liverpool, under Professor H. S. Hele-Shaw, LL.D., F.R.S., M.Inst.C.E., is set forth in an interesting pamphlet with plans and views of the main workshop and Walker engineering laboratories. Particulars are given of a number of valuable institutions. The examination for university scholarships takes place in May. Every scholar will be required to register as a student of the university, and to attend not less than four day lecture-courses in every term, unless the Senate expressly approve a smaller number of attendances.

### MR. BRADLEY'S SUCCESS.

I referred last month to the fact that an Institution of Mining and Metallurgy Research Scholarship of £50 had been awarded to Mr. R. S. Bradley. Mr. Bradley has been investigating the structure and behaviour of cast irons in the metallurgical laboratory of King's College, which is very well equipped for micro-photographic work. Although much is being done in this direction in the study of steel, very little attention has been paid to cast irons.

Mr. Bradley has recently established a laboratory at his father's foundry at Newark for chemical and micro-photographic examinations, and we understand he has already reaped considerable practical benefit from it.

### BIRMINGHAM UNIVERSITY ENGINEERING SOCIETY.

Reviewing the progress of this society during the past seven years, Professor F. W. Burstall, M.A., M.Inst.C.E., president, remarked that during the first session there were but few papers written by students ; even the discussions were, for the most part, confined to the older members of the society. Gradually, by slow development, the students took a larger and larger part not only in the papers, but in the government of the society. For his own part he considered the greatest thing that he had done to promote the usefulness of the society had been to abstain, as far as possible, from any interference with the methods of the students. He had always thought that for these college societies to be of real service in training students to think for themselves, the staff should not take a very prominent part in the government. Such treatment was likely to leave the students as helpless at the end of their course as they were in the beginning. The past three years had, he considered, shown a very great advance on the preceding time. The inauguration of the journal—the first of its kind in this country—was a sign that the Engineering Society was not content to stand still, but wished to advance in all directions which were likely to be of benefit to its members. At the present moment they were in a most prosperous condition, due almost entirely to the activity of those students responsible for conducting the affairs of the society and the journal.

### THE ROYAL COLLEGE OF SCIENCE.

The prospectus of the Royal College of Science and Royal School of Mines, makes a volume of 218 pages, packed with useful information. It is to be noted that application for the admission of students in the session commencing in October must be made on a specified form, and sent to the Registrar before the middle of June. In this form a statement is to be given of the studies which the applicant has already pursued, the examinations he has passed, and the names of a teacher or teachers to whom reference may be made. This application will be considered by the Dean and Council of the college, who will decide whether or not the candidates can be admitted.

Students must be free from organic disease and physical defect that would interfere with their studies, and a medical certificate to this effect may be required.

It is further pointed out that without some preliminary knowledge of mathematics, mechanics, chemistry, and physics, it is not possible for students to follow the courses with advantage. No fee-paying student will be entered for the Associateship Course unless he has obtained a First Class in the first stage of mathematics, or the first stage of practical mathematics ; and in the first stage of practical plane and solid geometry, theoretical mechanics (solids), theoretical mechanics (fluids), inorganic chemistry (theoretical), sound, light, and heat, and magnetism and electricity, or a pass in some higher stage of those subjects at the Examinations of the Board of Education, or can show to the satisfaction of the Council of the College, by having passed the Examinations of other recognised institutions or examining bodies, that he possesses the necessary elementary knowledge of those subjects, while for occasional students who desire only to take up certain specific branches of science some preliminary knowledge of such subjects will be required.

Admission is granted to persons desirous of attending certain courses of the lectures without the laboratory instruction, on payment of the lecture fees. Associates of the Royal College of Science or of the Royal School of Mines have the privilege of free admission to the Library and to all the courses of lectures.

### THE TREASURY GRANT TO UNIVERSITY COLLEGES.

A deputation introduced by Sir Oliver Lodge and representing the University Colleges of London, Manchester, Liverpool, Leeds, Birmingham, Nottingham, Bristol, Newcastle, Dundee and Sheffield, recently waited on the Chancellor of the Exchequer to point out the necessity of an increase in the present annual Treasury grant of £27,000 to University Colleges. We note that Mr. Austen Chamberlain will recommend that the vote be doubled for the present year, and that he hopes next year it may be increased to £100,000. It is obvious that the amount now set aside by the Treasury is quite inadequate for the maintenance of these institutions, and there are probably few directions in which the national funds could be better expended.

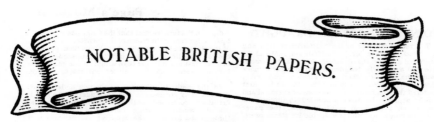

# NOTABLE BRITISH PAPERS.

A Monthly Review of the leading Papers read before the various Engineering and Technical Institutions of Great Britain.

## THE TREND OF MODERN INVENTION.

THE Junior Institution of Engineers received from their President Mr. J. Fletcher Moulton, K.C., F.R.S., M.P., a most instructive and suggestive address on "Invention." Incidentally he asked "In what direction is invention tending?" Here is the reply:—

In two directions which are well-nigh opposite or I ought rather to say complementary the one to the other. On the one hand the tendency is to divide manufacture up into many simple operations, each capable of being performed swiftly and well by a special machine designed solely for that purpose, and thus working under the most favourable circumstances for cheapness of production. Take for example the manufacture of machine-made watches—a manufacture which I am happy to say, is at last being vigorously taken up in England, after we have so long allowed ourselves to be distanced by our foreign competitors, both commercially and inventively. Each machine employed has only a minute operation for its share of the work, several being needed to perfect each piece. But these machines are so nearly automatic that the labour required for the most part needs no skill, and the rapidity with which they work makes the actual cost of production almost incredibly small, while the accuracy of the workmanship can be, and is, brought up to a pitch which completely satisfies all requirements of practical use. This tendency to subdivide the operations of the production until they are each capable of being performed either automatically or with unskilled labour, is having momentous effects in labour questions. Strangely to say, increase in accuracy of workmanship is tending to increase the demand for unskilled labour. The skill which used to be sought for in the workman is now embodied in the machine. This is due to what I may term the uniformity of mankind. The chief wants of each class are common to all the individuals that form it. Hence any rise in the standard of comfort of a nation produces a demand for millions of articles of one and the same kind, precisely such a demand as can best be satisfied by the unvarying but economical production of machines of the type of which I have spoken. I have no doubt that the growth of production as a whole will be so rapid that the total demand for skilled labour of all sorts in manufacture will not actually diminish, but I am equally sure that relatively to substantially unskilled labour, it will grow less and less. Rough and brutalising labour will no doubt be done away with, but its place will be taken by unskilled, rather than skilled labour.

For you must remember that precisely that same process of "coding" is being applied to bring each operation in production within the reach of unskilled labour. Take for example the machines that are used in domestic or trade life (such as typewriters, sewing machines, etc.), and which are produced in such vast numbers. Inventors are hard at work modifying the construction or configuration of each piece of these machines so as to lessen the cost of its production by enabling it to be made by some cheap process which dispenses with hand work and skilled labour. I have known a company in the United States, before they sold a single machine, spend two years and £20,000 in modifying the parts and their arrangement till each could be made by stamping or some similarly cheap method at a minimum of cost. And you must not forget that if wisely done this relegation to automatic machines and unskilled labour is an advantage to the public, because it brings with it as a consequence that absolute interchangeability of parts, which diminishes so vastly the cost of repairs.

Side by side with this tendency towards highly-specialised machines, each doing one small and separate operation, there is the other line of invention, i.e., of machines which combine—I would prefer to use the mathematical term, "integrate"—a whole series of successive operations and turn out a completed article. Here we find perhaps the greatest inventive triumph of our time. Take for instance the linotype: Type-setting, type-founding, and casting blocks of type as in stereotyping, had all been done by hand, and to some extent had been done mechanically, before the linotype came in. But it united them all in one machine, and enabled an operator, with little greater labour than in working a typewriter, to produce the set-up type cast in lines ready for printing.

It is difficult to say which class of machine is attracting most attention at the present moment—whether combination or division of operations, synthesis, or analysis, is taking the lead. If success is completely attained, the machine that combines in itself the whole series must always gain the day. But the penalty for falling short of perfection is heavier, and the danger greater. Each step brings its own liability to failure and the failure of a step has more serious and more far-reaching consequences. Yet we have abundant proofs on all sides of us that human ingenuity is equal even to this task. Here again the effects of the systematic pursuit of invention in the United States show themselves markedly. Few private inventors have, unaided, the means or the time to work out these complex problems. Mergenthaler, the inventor of the linotype, spent years of incessant labour before he came to a practical result. Plan after plan was devised by him only to be rejected, because the success it brought was too incomplete. At last he succeeded, and he and those who had supported him had their rich and deserved reward.

## POINTS ON FORCED DRAUGHT.

AT a recent meeting of the Staffordshire Iron and Steel Institute, Mr. R. B. Hodgson, A.M.I.M.E., contributed a paper, the object of which was to demonstrate that a properly designed forced draught furnace provides the means by which steam boilers can be most economically worked. To prove this, the author first considered the composition of fuel from its chemical and scientific aspect. He then proceeded to discuss the subject from a practical point of view, the paper containing many suggestions valuable for steam users. We quote the following :—

Under a good system of forced draught several advantages will follow :—

1. The working of the furnace will be entirely independent of both wind and weather.

2. There will be practically no loss through leaky brickwork.

3. The blast can be so regulated that no cold air need be drawn through the fire-doors when they are opened for cleaning, or otherwise attending to the fires, so that the heating surfaces may be kept at a more uniform temperature.

4. The firegrate may be reduced in area and the fire thereby concentrated, by which means the large excess of air over the theoretical quantity required for combustion, may be very materially reduced, with a resultant higher and more even temperature, and a consequent greater proportionate evaporation.

5. The heat of combustion will be available down to a much lower temperature than could be the case under natural draught. By the use of feed-water heaters, otherwise known as economisers, and increasing the number of economiser pipes, and by partially closing the dampers, the heated gases may be kept in contact with the boiler and its flues for a longer interval, and consequently enter the chimney much reduced in temperature.

6. The lower grade fuels, that can only be burned with difficulty under natural chimney draught, can be efficiently burned under forced draught.

7. Refuse fuels, such as coke dust, coal dust, pan breeze, spent tan, sawdust, ashpit refuse from puddling and reheating furnaces, and muffles, which it would be impossible to burn under chimney draught, may be burned with ease by forced draught, the higher rate of combustion compensating for the lower evaporating value of these fuels.

8. The evaporation under forced draught can be increased according to other conditions from 20 per cent. to 50 per cent. without loss of fuel efficiency.

### FORCED DRAUGHT IN OPERATION.

It is interesting to review the action of a boiler furnace working under an efficient system of forced draught. In this case the fire-door, which neither contains hole nor grid, is made an air-tight fit by being suitably machined. The ashpit is closed by a special casting, which carries a pair of cast-iron blower tubes, about 4 in. internal diameter. At the front end a steam nozzle is fixed, through which a jet of superheated steam issues ; this jet of steam causes a current of air to pass along the tube, and by this means any required air pressure may be maintained under the fire-bars. The fire-bars have narrow air spaces, from one-sixteenth inch to one-eighth inch apart. The bars themselves being narrow, a large number of air spaces are thereby provided, through which the air passes, the pressure in the ashpit forcing the air through these narrow air spaces. Should the fire-door be opened before the steam jets have been turned off, the flame will come out of the furnace towards the fireman, and it will be noticed there is no

" dip " in the furnace, but that the flame passes straight up from the fuel to the crown of the furnace. The action of the steam and air upon the bars keeps them perfectly cool, so that the clinker does not stick, consequently the fire-bars last much longer than would be the case if the furnaces were worked by either natural or induced draught.

In the methods of smoke prevention under forced draught, it is common to find instances of the grid in fire door combined with a closed ashpit, a bad practice which encourages currents of cold air to pass through the fire door the same as is the case with natural draught.

In the " Meldrum " system this difficulty is overcome by introducing a valvular dead-plate in the place of an ordinary dead-plate. When a furnace has been charged, and smoke is noticed issuing from the chimney, the valve is opened, and since this valve communicates with the ashpit, a thin stream of secondary air, at the pressure and temperature of the ashpit, is admitted over the fire at the front portion of the furnace tube, and has the effect of holding back green gases given off from the newly-fed fuel for a minute or so. Meanwhile, the full pressure of the forced draught is acting on the back portion of the furnace grate, consequently a brighter fire is generated and hotter gases are given off. So that when the valvular dead-plate is closed the green gases, in passing over the incandescent body of fire on the back of the grate, take fire, instead of passing away unconsumed. The valve may be opened to various widths according to the requirements of the furnace, a lever and rod being supplied for this purpose, and being in many instances automatic. That the smoke nuisance is overcome by these means is unquestionable. Some of the very worst known cases in this country have been cured in this way, and anyone who cares to take the trouble to test a furnace fitted up on these lines will find that a chimney giving off volumes of black smoke can be completely cured in about thirty seconds by opening the valve.

### CORRECT FORM OF BLOWER TUBES AND NOZZLES.

A word or two is needed with reference to the form of blowers, since so many people imagine that so long as a steam jet is inserted into a pipe the desired end is accomplished ; but this is only true if economy of steam and pressure of air are no object, and frequently steam jet blowers have been condemned through the wastefulness of badly-designed, cheap, and rubbishing apparatus. The proper form of blower tube is the outcome of long and careful experiments, some thousands of tests having been made with the blowers of all shapes and nozzles of all kinds, and the proportion of a modern blower and nozzle cannot be materially departed from without great loss of efficiency. With a correctly-designed and proportioned blower the amount of steam used is generally 3 per cent. of the total evaporation of the boiler.

### MEANING OF THE TERM "FORCED."

As some misunderstanding seems to have been created in the minds of many steam users, it will perhaps be as well if I mention two points and give the explanations. First—The term " Forced," as applied to boiler practice, is supposed to mean forcing the boiler. This fallacy is readily removed as one becomes familiar with the subject ; forcing a fire does not necessarily mean forcing the boiler. A natural draught furnace depends upon the difference in the specific gravity due to the difference in temperature between the columns of hot gases in the chimney and a corresponding column of the external air, consequently the draught must vary to some extent with the state of the atmosphere, humid or bright weather altering the conditions considerably. A strong wind blowing from the ashpit will sometimes necessitate the continual rousing of the fires to the discomfort of the fireman, as well as entailing a great waste of fuel. It must, therefore, be understood that forced draught is the term introduced to distinguish the draught provided by mechanical means—either steam jet, or fan to force air through the fire-bars—and not to force the boiler.

### AIR SUPPLY.

The second matter that does not appear to be clear to the steam user's mind, is regarding the air supply. Steam users naturally think forced draught gives a greater supply of air, on account of forced draught being stated as the means of accelerating combustion. This is a great mistake, forced draught given in a proper form provides less air than natural draught ; the necessary air for complete combustion is given at a greater velocity but through a smaller opening, because the ashpit is closed and the air supply comes through a pair of blower tubes. Taking for example a boiler with flues 30 in. in diameter, say under natural draught, we have the flue opening as ashpit; here we have, taking the two ashpits, an area of ashpit opening equal to 353.43 square inch, whereas the two 30-in. fines with forced draught will have four 4-in. tubes = 50.2656 square inches opening for the air passage, about 1 to 7. The air being carried up the tubes at a high velocity, creates a pressure in the closed ashpit, and the air under pressure escaping through the multitude of spaces between the fire-bars results in a thorough mixing of the air and the gases given off from the burning fuel. This is the principal reason why it is possible to work with so much less air supply under forced draught, than if natural or induced draught were used on the same boiler.

I have been continually asked by engineers and steam users whether boilers working with forced draught wear out more rapidly than boilers working under natural draught ? Many think that they do, and the idea is often engendered by the effects of crude attempts to use forced draught, and also by defective boiler design. This is undoubtedly a very erroneous idea, in fact the boot is upon the other leg. Experience proves that boilers are less injured by a properly applied forced draught, owing to the greater uniformity of temperature maintained in the boiler, furnaces, and combustion chambers, sudden alterations of heat and cold having most injurious effects on the interior parts of steam boilers. The principal way in which boilers are injured is by expansion and contraction of the fines, consequent upon the varying heat of the fire.

---

## TURBINES FOR LOW FALLS.

AT the ordinary meeting on Tuesday, 16th February, Sir William White, K.C.B., President, in the chair, the paper read was " The Forms of Turbines most suitable for Low Falls," by Alphonse Steiger, M.Inst.C.E. The following is an abstract of the Paper:—

The author draws attention in the first instance to the character of water powers with low falls. These are seldom constant, and in most cases both the fall and the water supply vary. The variations present a difficulty to the proper utilisation of such power, which, however, can be surmounted.

The prejudice against the utilisation of water power with low falls, which constitute the majority of water powers of the British Isles, arises principally out of the

disregard of their peculiarities and the want of knowledge of the results which may be obtained from a specially adapted turbine under varying conditions, and is traceable to many unsatisfactory installations of absolutely unsuitable turbines.

The author then considers the difference between impulse turbines and pressure turbines, showing by a diagram that varying the portion of the fall used for producing a pressure, that is, varying the degree of re-action, affords a means of adapting turbines to special requirements.

Pure impulse turbines are not altogether condemned for low falls, but places in which they are preferable to pressure turbines are extremely few in this country, and even the Haenel turbine, which is an intermediate type between the two, is not often suitable. An instance is given in order to show that such turbines, though excellent in themselves and efficient, do not necessarily make a satisfactory installation. The author then shows how the disadvantage of all axial-flow turbines, namely, the influence of the angles of vanes varying with the radial width, which is particularly serious in impulse turbines, is partially overcome.

### THE JONVAL TURBINE.

The parallel-flow pressure turbine, generally known as the Jonval turbine, is mentioned as particularly adapted to greatly fluctuating falls, under which a constant speed may be obtained without sacrifice of efficiency. Such a turbine is described, which the author has installed for a fall of 2 ft., and particulars of tests, made with similar turbines at the Zürich Waterworks, are given, to show that the part-gate efficiency of parallel-flow pressure turbines of this kind is extremely good.

It is characteristic of the large majority of re-action or pressure turbines that their part-gate efficiency is low, particularly so at less than half-gate, which renders them valueess for varying conditions ; but as in Europe the water-powers to be dealt with are chiefly of this kind, some European turbine-builders have succeeded in designing pressure turbines which are giving a very satisfactory efficiency even at quarter-gate.

### OTHER DEVELOPMENTS.

The demands of the generation of electricity by water-power, such as high speed, rapid regulation, and concentration of large power in one unit, have influenced very considerably the art of building turbines. In the first place, the desire for high speeds has led to a more general adoption of radial-flow turbines, of which the inward-flow type is preferable, being the more efficient.

An example of radial-outward-flow turbines is cited, giving the special reasons which have led to their adoption in one case of a relatively low fall, and the manner in which a quite satisfactory efficiency was obtained from this otherwise less efficient type. One new type, the cone turbine, is referred to as taking the place of the so-called "mixed-flow turbines," with a view to obtain a high speed under low falls, even for large units.

### CONSTRUCTION OF THE GATE.

One important factor in turbines is the construction of the gate, and it is shown in the paper that a distinction must be drawn between gates intended for adjustment of the turbine to a varying load, where a high part-gate efficiency is of less importance than rapid regulation, and gates for adapting a turbine to varying conditions, where a good part-gate efficiency is an essential condition.

Next, with special regard to electrical requirements, reference is made to the arrangement of several wheels on one common turbine shaft, vertical or horizontal, giving a few instances of the vertical arrangement which, under certain circumstances, is particularly advantageous, as the weight revolving on the footstep can be entirely balanced, and so the loss of power by friction is reduced almost to nil.

The necessity of placing the footstep of a turbine in an accessible position is alluded to, and the two kinds of overhead footsteps most commonly used are illustrated.

The paper concludes by drawing attention to the necessity of studying more closely the conditions of water-powers with low falls, and of paying greater attention to more careful adaptation of turbines to such falls, which would probably lead to a better appreciation of the water-powers of this country.

## TUNNELLING UNDER LAKE SUPERIOR.

AT a recent meeting of the Institution of Mining and Metallurgy, Mr. P. R. Robert, gave a most interesting account of some difficult tunnelling operations under Lake Superior, which were carried out in order to supply water to stamp mills in the Lake Superior copper region.

The stamping mills in this region generally contain from three to six heads each, and it is found that about thirty tons of water are required per ton of ore treated, the tendency nowadays being to increase rather than fall below that quantity. Owing to this demand for an abundant supply of water, the mills are located on the shore of the lake, though the disposal of the tailings is likewise a factor which has to be taken into serious consideration in determining the best position for them.

In the neighbourhood of the intake tunnel, described, there are five mills having twenty-one steam stamps with a daily crushing capacity of 9,500 tons.

In the case of the Adventure mill, the lower end of the mill and the pumping engine-house are 98 ft. from the lake. The lake shore faces north and is thus exposed to full force of the prevailing northerly winds and which are often very severe.

The shore consists of sandstone, which at the shore line rises to a height of 17 ft. above the lake. The lake bottom is also of sandstone and falls to the north with an inclination of about 2·6 in 100. The actual dip of the sandstone strata toward the north was not determined; it is probably about 20°. Along the shore there is much sand, but the lake bottom itself is nearly free from it. In winter time, when the wind is from the north, much floating ice is seen along shore, extending out for fully 100 ft., in hummocks as high as 20 ft. above lake level. This ice remains stationary until spring, when it melts away. In the fall, during storms, much vegetable matter is seen in suspension, such as leaves, water-logged wood, bark, etc., even as far out as 250 ft. from the shore.

The requirements of the mill were :—

(1) A full supply of water throughout the year.
(2) Absence of sand in the water.
(3) Possibility of increased supply in the event of the plant being enlarged.

To meet these requirements it was decided to sink a shaft on the shore, drive out a tunnel under the lake and let the water into the tunnel through holes bored into the bottom of the lake.

A shaft, 8 ft. by 10 ft., was started midway between the pump house and the lake, and sunk in the sandstone to a depth of 93 ft. The bottom of the shaft was made to converge to a point with the object of being able to extract by a sand pump any sand which might accumulate there.

The shaft was sunk by contract at $11 per ft. for labour and supplies.

At 81 ft. from the surface a tunnel 7 ft. by 5 ft. was started; it was driven by four men, two in a shift, using a Rand drill (No. 3 B) with the cylinder 3¼ in. diameter. The progress was, usually, about 150 ft. per month. The contract price for labour and supplies, not including tramming, was $5·50 per foot. The tunnel was driven for a total length of 887 ft. At first the amount of water percolating into it did not exceed fifty gallons per minute; but at a distance of 610 ft. a seam of white sandstone was cut which let in water freely. It appeared to be the prolongation of a similar seam which was cut in the shaft, 42 ft. below the surface.

After passing this seam the water in the drift increased considerably, until, at 825 ft., the inflow was 225 gallons per minute, and finally, at 887 ft., 350 gallons per minute. At this point a chamber was cut, 15 ft. long by 14 ft. wide and 9 ft. high.

The gradient of the tunnel was 6·76 in. per 100 ft. Immediately above the chamber the water was 22 ft. deep, leaving 25½ ft. between the bottom of the lake and the top of the chamber.

It was originally intended to drive the tunnel considerably further, to a point where the water would be deeper and the interval between the lake bottom and the tunnel less; but, in view of the extra expense and time which would have been required, it was eventually decided to stop the tunnel at the point already mentioned.

The writer intended to have ascertained the temperature of the rock at the extreme end of the tunnel, but the rapid inflow of the lake water prevented his doing it. On August 25th the temperature of the air in the end of the tunnel was 48° F.; at the same time the temperature of the water in the tunnel was 42° F.; whilst the temperature of the outside air was 74° F.

When the tunnel was completed boreholes had to be put down in the bed of the lake to let water into the chamber. The holes were bored by a Rand drill (Little Giant No. 7), with a cylinder 5½ in. in diameter. The drill was mounted on a frame standing on the deck of a scow or barge. The frame was similar to that of a pile-driving derrick, and projected over the side of the scow. The drill was arranged so as to slide in guides, and it could be raised or lowered by means of a small steam winch on the scow. Before the inlet holes were bored, four holes, each 2 in. in diameter and 4½ ft. deep, were bored in the lake bottom at points about 125 ft. from each corner of the scow when in position for drilling. In each of the holes was placed a 2 in. eye-bolt, and to each eye-bolt was attached a rope with wooden buoy. These anchor bolts were fixed by a diver and his helper, who had their own plant in a small scow, 18 ft. by 7 ft. by 2 ft. deep. This having been done, the bottom of the lake above the tunnel chamber was carefully cleared of small boulders, etc., and made as level as possible, partly by blasting. A steel plate ¼ in. thick was now bolted to the lake bottom immediately over the chamber; it was 14⅔ ft. by 11 ft., and in it twenty hole had been bored, 10½ in. in diameter and 36 in. centre to centre.

The services of the scow and a steam tug were now required. The scow, 20 ft. by 80 ft., carried on deck the drill and frame, a 40 h.p. boiler, a small steam winch, a feed pump, and a No. 5 Cameron steam pump to sluice out the holes. Near the corners the scow was provided with eight 8-in. spuds, with iron points, with the object of holding it in position while drilling was going on. The drill was worked by steam. The drill rods were made of steel of 1 in. diameter, in one piece. The drill bits were about 18 in. long, made with a socket into which the drill was fastened by a key.

The holes were started with a 9 in. drill and run down about 8 ft,. then an 8½-in. drill was substituted and the last 8 or 9 ft. were bored with an 8 in. drill.

As each hole was finished, a piece of pipe 8 in. in diameter internally and 8 ft. long was placed in it. 5 ft. of each pipe went into the hole, and 3 ft. projected above the plate on the lake bottom, each pipe being held in place by a flanged collar. The collar was fastened to the pipe by a set screw, and the flange was made wide enough to overlap the edges of the 10½-in. hole in the plate. The pipes were provided with strainers having holes 1 in. in diameter.

When drilling, the progress was generally about 2 ft. per hour. The seven last holes sunk in the most favourable season cost at the rate of $30·74 per foot.

This work was commenced so late in the season, and at the approach of winter, that the weather was very uncertain. Interruptions were therefore very frequent. The nearest port was ten miles distant, and very often it was necessary to run for shelter. On one occasion a storm came up so suddenly that the whole plant was very nearly lost, but work was proceeded with, as it was necessary to secure a water supply for two stamps during the winter. Only three holes were bored at that time, and it was estimated that these would furnish a sufficient supply under a head of 10 ft.

The pumping engine, when running to its full capacity at 75 revolutions per minute (48 in. stroke), would deliver 10,500 gallons per minute. With the three 8 in. holes, without the strainers, and two stamps in operation, the pumping engine made forty-five strokes per minute, and lowered the water in the shaft 10 ft.

The following spring seven more holes were drilled, making ten altogether. With three stamps in operation there was scarcely any perceptible lowering of the water in the shaft. The water was taken from the shaft to the pump by a suction pipe 36 in. in diameter, having a strainer with holes ⅞ in. in diameter. The total cost was $32,798·85.

The air for the drill in the tunnel was furnished by a small compressor on shore. There was also there a small steam hoist. Two steam pumps were used in sinking the shaft and driving the tunnel. The large Rand drill was the only machinery bought specially for the purpose, as nearly all the rest was borrowed from the mine and mill plants. The pumps were afterwards transferred to the mill where they were needed. The scow and tug were chartered.

## LIGHTSHIPS AND FOG-SIGNALS.

AT a meeting of the Institute of Marine Engineers an interesting lecture on "Lightships and Fog-Signals" was delivered by Mr. J. Sparling.

### EARLY LIGHTSHIPS.

Mr. Sparling opened his lecture by describing several of the well-known lightvessels that are moored around the coast. The first lightvessel, he said, was moored at the Nore Sand in 1732, and in 1736 another lightship was moored near the Dudgeon Shoal. The vessels were moored by hemp cables, which, owing to constant chafing, occasionally parted during the winter storms. The first lantern constructed to surround the mast of the lightvessel was designed by Mr. Robert Stevenson, the engineer of the Bell Rock lighthouse, in 1807.

Mr. Sparling described at some length the system of fog-signalling adopted on these vessels. The most powerful lights, he remarked, were unavailing in thick fog. In 1873-4 a series of experiments were carried out by the Trinity House, aided by Professor Tyndall and Sir James Douglass, the then engineer of the Trinity House. The information thus gained led to the adoption of sirens on a large scale, and reed signals in positions where a great range was required. Bells of two tons weight and Chinese gongs were at one time used on lightships, but these had been very largely superseded by manual reed horns. Bells were, however, still used at some stations.

At the Eddystone Lighthouse the two bells which were originally fixed on the gallery of the Douglass Tower had been replaced by the more effective gun cotton signal. At stations where reed and siren fog signals were fixed caloric engines of small power were installed, although in some recent instances these caloric engines had been superseded by oil engines of 20 b.h.p., capable of compressing 183 cubic feet of air per minute. With such an engine they employed two sirens of 5 in. diameter, each having a bent head and copper trumpet 6 ft. in diameter at the aperture, the sirens being geared together for the purpose of giving the same note.

The author considered that the fog-signals around our coasts were a matter on which the Corporation of the Trinity House had every reason to be congratulated. The experiments at St. Catherine's in 1901 fully justified the practice that had previously prevailed in the Trinity House service. In the course of the evening descriptions were given of the horns or trumpets designed by Lord Raleigh and others.

It was stated that, as a result of experiments that had been carried out, the Caskets Lighthouse had a 7 in. siren fitted with a 6 ft. mushroom trumpet, whilst the Nash and Whitby Lighthouses were equipped with two 5 in. sirens fitted with Raleigh trumpets.

Many excellent lantern slides were shown by the lecturer, who was accorded a hearty vote of thanks, on the motion of Mr. K. C. Bales.

# BOOKS OF THE MONTH.

### "THE ADJUSTMENT OF WAGES."

A Study in the Coal and Iron Industries of Great Britain and America. By W. J. Ashley. With four maps. Longmans, Green and Co. 12s. 6d. net.

Intended as a preliminary survey in a very wide field of investigation, these lectures should afford the student an insight into the problems which underlie the adjustment of wages, and, incidentally, the constitution and working of conciliation boards. The coal, iron, and steel industries have been chosen for consideration, primarily on account of their magnitude and the fact that the persons engaged in them number somewhere between an eighth and a ninth of the income-earning population of Great Britain. Moreover, as the largest of industries, they serve to exhibit the complete substitution of collective or corporate bargaining about wages for individual agreement. The lion's share of attention falls to the coal industry, and in Lecture II. some interesting deductions are drawn from a consideration of the various British Boards of Conciliation. An instructive chapter on "Prices and Wages," in which the *raison d'etre* of the sliding scale is carefully explained, is followed by a consideration of general rates and their interpretation, the hours of labour, etc. The fifth lecture is devoted to the American Coal Fields, and Joint Agreements, and a large share of attention is given to the anthracite problem, the author going very fully into the conditions underlying the recent American coal strike. He then proceeds to discuss the network of Boards of Conciliation which cover the whole field of the iron industry, and a large part of the field of the steel industry in Great Britain. It is remarked that all these boards depend for their efficacy on the existence and practical recognition by the employers of strong trades' unions among the men, while throughout the manufactured iron trade (with the exception of the iron founders) as well as in several branches of the steel trade, general rates of wages are determined by sliding scales. That the sliding scale method seems to possess so much vitality in the iron trade, while it has had to be abandoned in name and greatly modified in practice in the coal mining industry, is explained by two circumstances. "In the first place, the 'long-contract' or 'selling-ahead' system, which has always endangered the sliding scale in the coal industry, does not exist at all in the iron and steel trades." "In the second place, there is some reason to believe that combination among the employers has been more successful in 'regulating' prices in the iron than in the coal industry."

Discussing the iron industry of America, the author remarks that American business men are in their way industrial statesmen :—"They will not be content simply to suppress unions again and again ; they will seek an alternative policy. Mr. Schwab's own policy has just been announced as one of profit-sharing ; and he is now carrying through a scheme to facilitate the acquisition of stock in the Steel Corporation by the operatives employed. I am ready to confess that I am not absolutely sure he may not succeed thereby in killing unionism in his particular branch of trade. I hold no brief for unionism. It is not an end in itself, but only the means to an end ; and the end or ends may conceivably be reached in other ways. But I doubt whether the management of the Steel Corporation realises the grave practical difficulties in the way of the permanent success of this alternative policy. The problem is one very largely of business administration—how to deal, with least expenditure of time and energy, with large masses of men engaged in more or less similar work. And I should not be surprised if American business administrators, after desperately fighting unions until they are tired, should suddenly come to terms with the men, and organise 'joint agreements,' with a completeness and with a thoroughness the world has never seen. The critic may object to such an alliance, and talk of the interests of the consumer. But we have so long idolised the consumer, and with such unsatisfactory results, that society may do well to try the experiment of thinking first of the producers—of all the producers."

The final lecture is concerned with the legal position of trades unions. Incidentally, the author remarks that "the law, as stated in the Taff Vale case, is pretty certain to remain for the future ; and unions will in future be responsible for the wrongful acts of their representatives. The real question which demands an answer is, What are wrongful acts ? I shall not attempt," he continues, "to disentangle the complexity of the present legal position as formulated in the decisions of the last three or four years. Eminent lawyers are still capable, in the interest of their clients, of making large and positive assertions in court ; but they know very well that in reality the law is at present in a state of great obscurity and uncertainty."

A large section of the volume consists of appendices, invaluable to the student, and comprising rules of Conciliation Boards, joint agreements, report and awards of the Anthracite Coal Strike Commission, etc., etc.

We do not desire to quarrel with the matter or manner of Professor Ashley's volume, but it has occurred to us that in a future edition it may be found desirable to make the work easier of consultation for those who desire to use it as a reference volume. There are four excellent maps dealing with the English and American coalfields.

---

*A number of reviews are unavoidably held over.*

# OUR DIARY.

## January.

**22nd.**—The Transvaal mineral output for last month included—silver, 31,406 oz. ; gold, 288,824 oz.; diamonds, 29,700 carats ; coal, 192,784 tons.

**23rd.**—Lord Inverclyde presides at the dinner of the Institution of Engineers and Shipbuilders in Glasgow.—About 70 per cent. of the white male inhabitants of the Transvaal have signed the petition, which closes to-day, asking the Government to pass law for the importation of Asiatic unskilled labour.

**26th.**—The general revenue of Cape Colony continues to show a heavy decrease.—The Transvaal Labour Importation Ordinance passes through Committee in the Legislative Council.

**28th.**—Sir William White, at the annual dinner of the Liverpool Engineering Society, advocates the teaching of naval architecture in the Liverpool University.

**29th.**—Half-yearly meeting of the Conciliation Board in the North of England iron and steel trade at Newcastle-on-Tyne.

**30th.**—Contracts for the construction of two battleships for Japanese Navy, signed in London by representatives of the Mikado's Government—one of these is to be built by Messrs. Vickers, Sons and Maxim, the other by Sir W. G. Armstrong, Whitworth and Co.

## February.

**1st.**—Issue of Blue Book on the use of Electricity in Mines.—Mr. David B. Butler, president of the Society of Engineers, delivers his inaugural address at the Royal United Service Institution.—The Lancashire and Yorkshire Railway Company introduce a new series of powerful tank locomotives.

**2nd.**—A discussion takes place at the meeting of the London County Council with regard to precautions against theatre fires—it is stated that London managers agree to submit their stage fittings to a process which will render them non-inflammable.

**4th.**—Launch of the battleship *New Zealand*.—The Works Committee of the Mersey Docks and Harbour Board recommend the expenditure of £295,000 for improvements.

**6th.**—Japan breaks off diplomatic relations with Russia.

**7th.**—Collapse of a portion of Dove Holes Tunnel on the Midland main line, attributed to the continuous wet weather.—The Federal Steamship Company's tender accepted for a monthly steamship service between New Zealand and the West of England.—Disastrous fire at Baltimore causes loss of employment to fifty thousand people.

**8th.**—Postponement of the opening of the Great Northern and City Railway, owing to alterations necessary in connection with the signalling arrangements.—First act of war between Japan and Russia.

**9th.**—In consequence of communications received from the home Government, the Legislative Council in Pretoria decide to postpone taking any steps with regard to the importation of native labour.—The London County Council accept the tender of Messrs. Price and Reeves, London, for the construction of the Rotherhithe Tunnel, amounting to £1,088,484.—An expenditure not exceeding £181,400 for the construction of the fourth section of the enlargement of the northern outfall sewer is sanctioned.

**10th.**—Lord Carrington presides at a meeting in the Queen's Hall "to protest against the importation of indentured Chinese labour into the Transvaal."—A meeting of persons interested in the Port of London pass a resolution in favour of a scheme for the construction of a barrage across the Thames at Gravesend.

**11th.**—Wreck of the steamship *Yeoman* at Corcubion.

**12th.**—Opening of the Automobile Club Exhibition at the Crystal Palace.—The Legislative Council, Pretoria, gives its assent to the Labour Importation Ordinance.—The revenue of the Transvaal for the half-year ending December last amounts to £2,105,062, and the expenditure £2,253,428.—Death of Mr. William Thomas Ansell.—During the week the floating graving dock, illustrated in our November issue, arrived safely at Durban.

**13th.**—The Engineering Standards Committee issue a statement of the work now in progress under its auspices.—Junior Institution of Engineers hold their annual dinner.

**14th.**—Opening of the Great Northern and City Railway.

**16th.**—Ratification of the combination of Scotch steelmakers to maintain prices on a profitable basis.

**17th.**—An amendment lost in the House of Commons condemning the introduction of Chinese labour into the Transvaal.—Leading members of the commercial community in Johannesburg declare that the slackness of trade is universal, and that unless labour is supplied to the mines further restriction is inevitable, which would mean disaster.

**18th.**—M. Curie delivers a lecture on radium at the Sorbonne.—An explosion of nitro-glycerine occurs at Messrs. Curtis and Harvey's premises at Cliffe.

**19th.**—The Civil and Mechanical Engineers Society inspect the tramway tunnel which is being constructed by the London County Council under Kingsway.—The Labour Ordinance promulgated in the Transvaal, but awaits the Royal approval.

**20th.**—His Majesty the King inspects the Royal Naval College, Osborne, and the Naval Barracks, Portsmouth.—Deposits of tin ore and excellent coal found in Lower Burma.—Opening of the London County Council pumping station, Lot's Road, Chelsea.

# NEW CATALOGUES AND TRADE PUBLICATIONS.

**John Lang and Sons,** Johnstone, near Glasgow, forward an interesting catalogue descriptive of various lathes fitted with Lang's Patent Variable Speed Drive and Automatic Speed-Changing Mechanism. With the former, in sliding, surfacing, and screw-cutting lathes, the correct surface speed may be had for every diameter, however small the variation. With both, in surfacing and boring lathes while surfacing work, the revolutions of spindle automatically increase as the diameter being turned becomes smaller, thus keeping the tool cutting at practically a constant surface speed. It is claimed that these machines constitute a triumph in lathe design.

**Chambers, Scott and Co., Motherwell, N.B.**—From this company we have received an artistic catalogue of Electric Overhead Travelling Cranes (E.C. No. 1904). The special features of these cranes are set forth as follows : All bearings being bushed, alignment and adjustment are assured. The Standard Crane Crabs have no overhung wheels, and all the gearing is placed between the cheeks, making a very satisfactory arrangement, and giving a neat and compact trolley. In the designs the number of gear reductions is kept as small as possible, and accessibility to all parts and simplicity of construction is aimed at.

**Quaker City Rubber Company, 101, Leadenhall Street, E.C.**—This company, whose managers for Great Britain are Messrs. Ronald Trist and Co., have issued a pamphlet containing much useful information on the subject of Daniel's P.P.P. Packing. The advantages of this packing have already been referred to in PAGE'S MAGAZINE. We recommend steam users to obtain the pamphlet and investigate the matter for themselves.

**Pethick Bros., London and Plymouth.**—This firm issues a well-illustrated booklet, with the significant title, "Enough Granite to Build Ten Londons." This is based on the statement of an engineer who was called upon to inspect the Princetown Quarries of Dartmoor, and to report as to their resources. The pamphlet has some admirable illustrations of the principal buildings, works, etc., for which Devonshire granite has been used, all of which are of the greatest interest to engineers.

**Lancaster and Tonge, Ltd.,** are issuing a fourth edition of their well-known "Slate" catalogue. In order to meet the demand for a more compact steam trap, a special 1903 pattern has been designed, and it is prominently illustrated in this catalogue. To show the compactness of the new trap, it may be mentioned that the ¾-in. trap is 9 in. high, 10 in. long, and 5½ in. wide over all. An interesting woodcut shows the cylinders of a compound engine which is fitted with the "Lancaster" metallic packings, pistons, piston valves, steam dryer, and steam traps (1903 patent). The firm is also issuing a pamphlet entitled the "Lancaster" Specialities. It includes many highly favourable opinions from users of the "Lancaster" metallic packings.

**W. N. Brunton and Son,** of Musselburgh, Scotland, forward a price list of Galvanised Arc Lamp Ropes. It is claimed for these ropes that they are many times

more durable than Manilla, and that they neither stretch nor shrink.

**C. W. Hunt Company.**—From this company we have received a well-illustrated booklet on the Hunt Industrial Railway. In addition to a large number of illustrations showing the varied types of cars, standard and special, built by the company, the booklet contains much valuable information relating to the Industrial Railway not to be found elsewhere.

**John Kaye and Sons, Ltd,** have issued a new price list for 1904, showing the latest improvements in their oil cans. As all previous price lists are cancelled, the new issue should be obtained and filed for reference.

**Thos. Thompson and Co., 35, New Broad Street, E.C.,** forward a booklet description of Thompson's "Standard Tachometers, with numerous illustrations. The special feature of Amund's new Tachometers are thus set forth : all sliding friction done away with ; no oiling of inner mechanism needed ; slightest possible wear ; extreme sensitiveness even under the shortest and slightest variation of speed ; constant reliability ; simple, self-contained, and strong design ; absence of delicate parts. We note that the firm acts as agents for the Edwards Air Pump Syndicate, Ltd.

**Trading and Manufacturing Company, Ltd.,** are issuing a four-page illustrated description of the T. and M. Individual File, the object of which is to bind loose sheets or pamphlets in readily accessible book form.

**The Cling-Surface Manufacturing Company,** of Buffalo, U.S.A., send us a breezy pamphlet on Cling-Surface and Belt Management. This is a production which deserves to be read, if only for the fact that it is well arranged and gets straight to the point.

**Empire Typewriter Company, Ltd.**—The latest booklet of this Company should be seen by prospective purchasers of typewriters. We note that the price of the regular foolscap model is £12 12s. Irrespective of price the machine has the distinct advantage of visible writing.

**Calendars.**—We have to acknowledge the receipt of ornamental and useful sheet wall calendars from the Whitman and Barnes Manufacturing Company, of 149, Queen Victoria Street, and W. H. Willcox and Co., Ltd., of 23, Southwark Street, S.E.

---

Messrs. Lepard and Smiths, Ltd., of 29, King Street Covent Garden, W.C., forward some samples of typewriting paper in white and five tints, with the useful suggestion that these may be identified with different business departments. The paper is certainly pleasant to the eye and seems admirably adapted for its purpose ; moreover, we note that it is of British manufacture.

---

The Worthington Pump Company, Ltd., write that the author of our articles on Birmingham University was under a misapprehension in stating that the cooling tower for circulating water supplied by them was built by a competing firm.

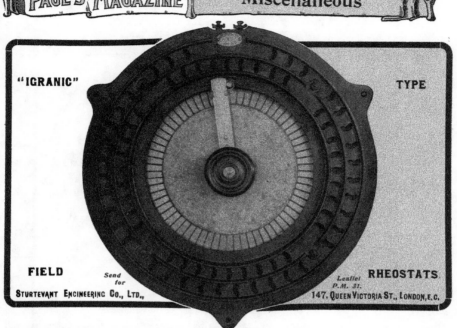
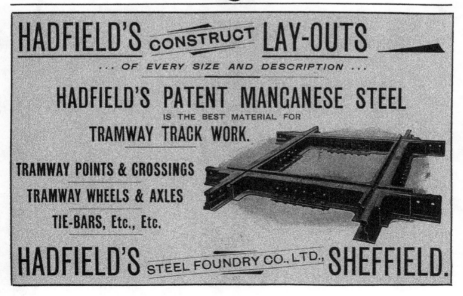
D

# JOHN FOWLER & CO.

## (LEEDS) LIMITED.

**Electrical and General Engineers.**

Steam Plough Works:
LEEDS.

Fowler's Road Locomotive. Designed for all kinds of Steam
Haulage, and is also available for temporary belt driving.
Three sizes of this Engine are standardized, and employed
approximately for 20, 30, and 40 ton loads. A special heavy
Engine is also made equal to a load of 50 tons, and called
the "Lion" type. The Engine was thus named by the
War Office Authorities, who employed a number of them
in the South African Campaign.

# GALLOWAYS L<sup>TD.</sup>,

## MANCHESTER.

## High=Speed Engines.

QUICK DELIVERY.

### Compound and Triple Expansion.

### SLOW=SPEED CORLISS ENGINES for Mill Work.

## Galloway Boilers.

IMMEDIATE DELIVERY.

### All Sizes and all Pressures.

---

**WROUGHT STEEL SUPERHEATERS.**    The Safest and Most Reliable on the Market.

Telegrams: GALLOWAY, MANCHESTER.     London Office: 17, PHILPOT LANE, E.C.

54

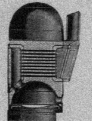

# The Mirrlees-Watson Co.,
### LIMITED,
## GLASGOW, SCOTLAND.

### CONDENSING PLANT FOR HIGH VACUUM.

LONDON OFFICE:
158, Gresham House,
Old Broad Street. E.C.

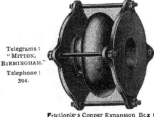

59

BENNIS STOKERS, CONVEYING AND ELEVATING PLANT, ON CORNISH, LANCASHIRE, AND WATER-TUBE BOILERS.

FOR
**Gas Engines.**

FOR
**Furnace Work.**

# GAS

W. F. MASON, Ltd.,
*Engineers &
Contractors,*
MANCHESTER.

# PLANTS
(BITUMINOUS, COKE, OR ANTHRACITE.)

**FURNACE WORK OF ALL KINDS IS OUR SPECIALTY.**

# MELDRUM·BROS·LTD

### Timperley near Manchester.

The Meldrum Destructor effectually disposes of refuse once and for all. Not merely eliminating the undesirable, it conserves the valuable residue, and proves at once an indefatigable scavenger and money-maker. Wideawake corporations are now using the surplus heat for steam power in their electric light and tram departments, sewage pumping, and other municipal enterprises.

Engineers and others interested in the subject of Contemplated Destructor Installations, are invited to communicate with us.

This illustration shows half the plant destined for Johannesburg erected in our shop before being shipped.

**TIMPERLEY, MANCHESTER,**
AND
**66, VICTORIA STREET,**
**WESTMINSTER,**
        **LONDON.**

# THE "HORSFALL" DESTRUCTOR — Hot Blast.

*Adopted at*

Westminster,
Fulham,
Brussels,
Leeds,
Bradford,
Hull,
Sheffield,
Monaco,
&c., &c.

*Adopted at*

Oldham,
Finsbury,
Accrington,
Blackpool,
West Hartlepool,
Leamington,
Hamburg,
Windsor,
&c., &c.

Plants at work . . . . . . **60**
Contracts in hand . . . . . **17**
New Orders since Jan. 1st, 1903 . **16**

*Including DURBAN, BLOEMFONTEIN, CAPETOWN (Hospital), FOLKESTONE, SWANSEA, &c., &c.*

# The Horsfall Destructor Co., Ltd.

Lord Street Works, LEEDS.

Telegrams: "DESTRUCTOR, LEEDS."
A B C, 5th Edition, and LIEBER'S CODES.

19, Old Queen Street,
Westminster, LONDON, S.W.

Telegrams: "DESTRUCTOR, LONDON."

E

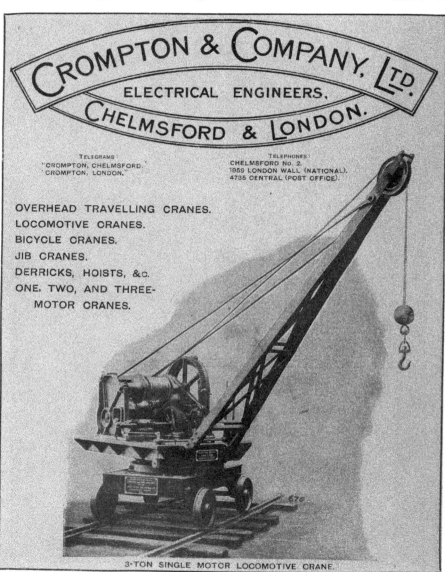

# CROMPTON & COMPANY, LTD.

## ELECTRICAL ENGINEERS,

## CHELMSFORD & LONDON.

TELEGRAMS:
"CROMPTON, CHELMSFORD."
"CROMPTON, LONDON."

TELEPHONES:
CHELMSFORD No. 2.
1959 LONDON WALL (NATIONAL).
4735 CENTRAL (POST OFFICE).

OVERHEAD TRAVELLING CRANES.

LOCOMOTIVE CRANES.

BICYCLE CRANES.

JIB CRANES.

DERRICKS, HOISTS, &c.

ONE, TWO, AND THREE-
    MOTOR CRANES.

3-TON SINGLE MOTOR LOCOMOTIVE CRANE.

# JOSEPH BOOTH & BROS.

## LTD.,

## RODLEY,

## LEEDS.

For

Lifting

Machinery,

&c.

40-ton Steam Goliath Crane at the new L. & N. W. Railway Goods Yard, Sheffield.
And also supplied to Midland, Lancashire & Yorkshire, and Great Western Rys., &c.

Cranes, Winding Engines,
Overhead Travellers of
Every Description, Driven
by Steam, Electricity, or
·Hydraulic Power.

London Agents:
A. E. W. GWYN, Ltd.,
75a, Queen Victoria St., E.C.

Telephone:
20 STANNINGLEY.

Telegrams:
·· "CRANES, RODLEY."
"ASUNDER, LONDON."

As supplied to Crown Agents for the Colonies and Government Departments.

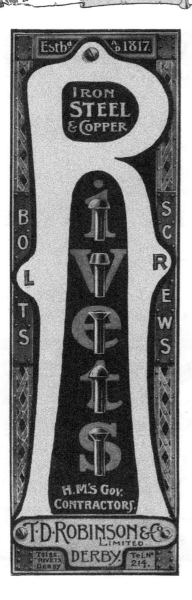

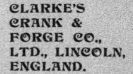

# Farnley Iron

HAMMERING BLOOMS.

Farnley **Bar Iron** is used in **Mining** for pit cages, suspending gear, and other important parts, and on all the leading **Railways** in Great Britain, India, and the Colonies, for shackles and other vital parts subjected to repeated shocks.

Farnley Iron will stretch cold from $1\frac{1}{8}$ in. to $2\frac{1}{8}$ in. in a length of **6** in. before fracture, and is safest for **welding**.

*Address:* **The Farnley Iron Co., Ltd., Leeds, England.**

HEAD OFFICE —
ST PAUL'S SQUARE,
BIRMINGHAM.

WATERLOO CHAMBERS
19, WATERLOO STREET,
GLASGOW.

# SAMᴸ BUCKLEY
★ ★ ★
## STYRIAN STEEL WORKS
## SHEFFIELD

BRAND BÖHLER

CORRESPONDENCE SOLICITED.    PROMPT REPLIES.    PROMPT DELIVERIES

BRAND
STYRIAN    STEEL.
SAMᴸ BUCKLEY    SHEFFIELD
BOHLER

BRAND    BOHLER
# MANUFACTURERS
★ ★ ★
# ROLLERS AND FORGEMEN
OF EVERY DESCRIPTION OF
# CRUCIBLE CAST & MILD STEELS
CORRESPONDENCE SOLICITED.    PROMPT REPLIES.    PROMPT DELIVERIES

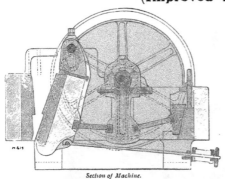

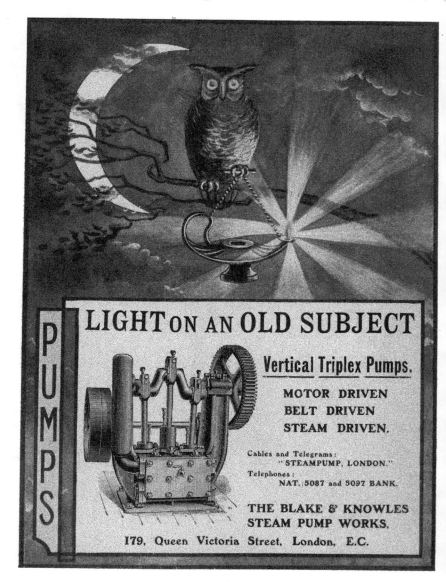

86

# JOHN STIRK & SONS
## HALIFAX.
(Established 1866.)

Electric-driven Horizontal Boring Machine, with 7 in. spindle, differential feeds, and universal chuck for bars.

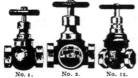

Printed for the Proprietors by SOUTHWOOD, SMITH & Co., Limited, 6, 7, 8, 9, Plough Court, Fetter Lane, London, E.C., and Published at Clun House, Surrey Street, Strand, London, W.C.

CPSIA information can be obtained
at www.ICGtesting.com
Printed in the USA
BVHW081831061118
532319BV00012B/1234/P